At the Deathbed of Darwinism

by Eberhard Dennert

Authorized Translation

By E. V. O'HARRA and JOHN H. PESCHGES

CONTENTS

PREFACE 9

INTRODUCTION 27

CHAPTER I.

--The Return to Wigand--The Botanist, Julius von Sachs--The Vienna Zoologist, Dr. Schneider

CHAPTER II.

--Professor Goethe on "The Present Status of Darwinism"--Explains the Reluctance of certain men of Science to Discard Darwinism

CHAPTER III.

--Professor Korchinsky Rejects Darwinism--His Theory of Heterogenesis--Professor Haberlandt of Graz--Demonstration of a "Vital Force"--Its Nature--The Sudden Origination of a New Organ--Importance of the Experiment.

CHAPTER IV.

--Testimony of a Palaeontologist, Professor Steinmann--On Haeckel's Family Trees--The Principle of Multiple Origin--Extinction of the Saurians--"Darwinism Not the Alpha and Omega of the Doctrine of Descent"--Steinmann's Conclusions

CHAPTER V.

--Eimer's Theory of Organic Growth--Definite Lines of Development--Rejects Darwin's Theory of Fluctuating Variations--Opposes Weismann--Repudiates Darwinian "Mimicry"--Discards the "Romantic" Hypothesis of Sexual Selection--"Transmutation is a Physiological Process, a Phyletic Growth"

CHAPTER VI.

--Admissions of a Darwinian--Professor von Wagner's Explanation of the Decay of Darwinism--Darwinism Rejects the Inductive Method, Hence Unscientific--Wagner's Contradictory Assertions

CHAPTER VII.

--Haeckel's Latest Production--His Extreme Modesty--Reception of the Weltraetsel--Schmidt's Apologia--The Romanes Incident--Men of Science Who Convicted Haeckel of Deliberate Fraud

CHAPTER VIII.

--Grottewitz Writes on "Darwinian Myths"--Darwinism Incapable of Scientific Proof--"The Principle of Gradual Development Certainly Untenable"--"Darwin's Theory of "Chance" a Myth"

CHAPTER IX.

--Professor Fleischmann of Erlangen--Doctrine of Descent Not Substantiated--Missing Links--"Collapse of Haeckel's Theory"--Descent Hypothesis "Antiquated"--Fleischmann Formerly a Darwinian--Haeckel's Disreputable Methods of Defense

CHAPTER X.

--Hertwig, the Berlin Anatomist, Protests Against the Materialistic View of Life"--No Empiric Proof of Darwinism--"The Impotence of Natural Selection"--Rejects Haeckel's "Biogenetic Law"

CONCLUSION.--Darwinism Abandoned by Men of Science--Supplanted by a Theory in Harmony With Theistic Principles

PREFACE.

The general tendency of recent scientific literature dealing with the problem of organic evolution may fairly be characterized as distinctly and prevailingly unfavorable to the Darwinian theory of Natural Selection. In the series of chapters herewith offered for the first time to English readers, Dr. Dennert has brought together testimonies which leave no room for doubt about the decadence of the Darwinian theory in the highest scientific circles in Germany. And outside of Germany the same sentiment is shared generally by the leaders of scientific thought. That the popularizers of evolutionary conceptions have any anti-Darwinian tendencies cannot, of course, be for a moment maintained. For who would undertake to popularize what is not novel or striking? But a study of the best scientific literature reveals the fact that the attitude assumed by one of our foremost American zoologists, Professor Thomas Hunt Morgan, in his recent work on "Evolution and Adaptation," is far more general among the leading men of science than is popularly supposed. Professor Morgan's position may be stated thus: He adheres to the general theory of Descent, i.e., he believes the simplest explanation which has yet been offered of the structural similarities between species within the same group, is the hypothesis of a common descent from a parent species. But he emphatically rejects the notion--and this is the quintessence of Darwinism--that the dissimilarities between species have been brought about by the purely mechanical agency of natural selection.

To find out what, precisely, Darwin meant by the term "natural selection" let us turn for a moment, to his great work, The Origin of Species by Means of Natural Selection. In the second chapter of that work, Darwin observes that small "fortuitous" variations in individual organisms, though of small interest to the systematist, are of the "highest importance" for his theory, since these minute variations often confer on the possessor of them, some advantage over his fellows in the quest for the necessaries of life. Thus these chance individual variations become the "first steps" towards slight varieties, which, in turn, lead to sub-species, and, finally, to species. Varieties, in fact, are "incipient species." Hence, small "fortuitous" fluctuating, individual variations--i.e., those which chance to occur without predetermined direction--are the "first-steps" in the origin of species. This is the first element in the Darwinian theory.

In the third chapter of the same work we read: "It has been seen in the last chapter that amongst organic beings in a state of nature there is some individual variability. * * * But the mere existence of individual variability and of some few well-marked varieties, though necessary as a foundation of the work, helps us but little in understanding how species arise in nature. How have all those exquisite adaptations of one part of the organization to another part, and to the conditions of life, and of one organic being to another being, been perfected? * * *" Again it may be asked, how is it that varieties, which I have called incipient species, become ultimately converted into good and distinct species, which in most cases obviously differ from each other far more than do the varieties of the same species? How do those groups of species which constitute what are called distinct genera arise? All of these results follow from the struggle for life. Owing to this struggle, variations, however slight and from whatever cause proceeding, if they be in any degree profitable to the individuals of a species, in their infinitely complex relations to other organic beings, and to their physical conditions of life, will tend to the preservation of such individuals and will generally be inherited by the offspring. The offspring also will thus have a better chance of surviving, for of the many individuals of any species which are periodically born, but a small number can survive. I have called this principle by which each slight variation, if useful, is preserved, by the term, "natural selection." Mr. Darwin adds that his meaning would be more accurately expressed by a phrase of Mr. Spencer's coinage, "Survival of the Fittest."

It may be observed that neither "natural selection" nor "survival of the fittest" gives very accurate expression to the idea which Darwin seems to wish to convey. Natural selection is at best a metaphorical description of a process, and "survival of the fittest" describes the result of that process. Nor shall we find the moving principle of evolution in individual variability unless we choose to regard chance as an efficient agency. Consequently, the only efficient principle conceivably connected with the process is the "struggle for existence;" and even this has only a purely negative function in the origination of species or of adaptations. For, the "surviving fittest" owe nothing more to the struggle for existence than our pensioned veterans owe to the death-dealing bullets which did not hit them. Mr. Darwin has, however, obviated all difficulty regarding precision of terms by the remark that he intended to use his most important term, "struggle for existence" in "a large and metaphorical sense."

We have now seen the second element of Darwinism, namely, the "struggle for life." The theory of natural selection, then, postulates the accumulation of minute "fortuitions" individual modifications, which are useful to the possessor of them, by means of a struggle for life of such a sanguinary nature and of such enormous proportions as to result in the destruction of the overwhelming majority of adult individuals. These are the correlative factors in the process of natural selection.

In view of the popular identification of Darwinism with the doctrine of evolution, on the one hand, and with the theory of struggle for life, on the other hand, it is necessary to insist on the Darwinian conception of small, fluctuating, useful variations as the "first-steps" in the evolutionary process. For, this conception distinguishes Darwinism from the more recent evolutionary theory, e.g., of De Vries who rejects the notion that species have originated by the accumulation of fluctuating variations; and it is quite as essential to the Darwinian theory of natural selection as is the "struggle for life." It is, in fact, an integral element in the selection theory.

The attitude of science towards Darwinism may, therefore, be conveniently summarized in its answer to the following questions:

1. Is there any evidence that such a struggle for life among mature forms, as Darwin postulates, actually occurs?

2. Can the origin of adaptive structures be explained on the ground of their utility in this struggle, i.e., is it certain or even probable that the organism would have perished, had it lacked the particular adaptation in its present degree of perfection? On the contrary, is there not convincing proof that many, and presumably most, adaptations cannot be thus accounted for?

The above questions are concerned with "the struggle for life." Those which follow have to do with the problem of variations.

3. Is there any reason to believe that new species may originate by the accumulation of fluctuating individual variations?

4. Does the evidence of the geological record--which, as Huxley observed, is the only direct evidence that can be had in the question of evolution--does this

evidence tell for or against the origin of existing species from earlier ones by means of minute gradual modifications?

We must be content here with the briefest outline of the reply of science to these inquiries.

1. Darwin invites his readers to "keep steadily in mind that each organic being is striving to increase in geometrical ratio." If this tendency were to continue unchecked, the progeny of living beings would soon be unable to find standing room. Indeed, the very bacteria would quickly convert every vestige of organic matter on earth into their own substance. For has not Cohn estimated that the offspring of a single bacterium, at its ordinary rate of increase under favorable conditions, would in three days amount to 4,772 billions of individuals with an aggregate weight of seven thousand five hundred tons? And the 19,000,000 elephants which, according to Darwin, should to-day perpetuate the lives of each pair that mated in the twelfth century--surely these would be a "magna pars" in the sanguinary contest. When the imagination views these and similar figures, and places in contrast to this multitude of living beings, the limited supply of nourishment, the comparison of nature with a huge slaughterhouse seems tame enough. But reason, not imagination, as Darwin observes more than once, should be our guide in a scientific inquiry.

It is observed on careful reflection that Darwin's theory is endangered by an extremely large disturbing element, viz., accidental destruction. Under this term we include all the destruction of life which occurs in utter indifference to the presence or absence of any individual variations from the parent form. Indeed, the greatest destruction takes place among immature forms before any variation from the parent stock is discernible at all. In this connection we may instance the vast amount of eggs and seeds destroyed annually irrespective of any adaptive advantage that would be possessed by the matured form. And the countless forms in every stage of individual development which meet destruction through "accidental causes which would not be in the least degree mitigated by certain changes of structure or of constitution which would otherwise be beneficial to the species." This difficulty, Darwin himself recognized. But he was of opinion that if even "one-hundredth or one-thousandth part" of organic beings escaped this fortuitous destruction, there would supervene among the survivors a struggle for life sufficiently destructive to satisfy his theory. This suggestion, however, fails to meet the difficulty. For, as Professor Morgan points out, Darwin assumes "that a second competition takes place after the first destruction of individuals has occurred, and this presupposes that more individuals reach maturity than there is room for in the economy of nature." It presupposes that the vast majority of forms that survive accidental destruction, succumb in the second struggle for life in which the determining factor is some slight individual variation, e.g., a little longer neck in the case of the giraffe, or a wing shorter than usual in the case of an insect on an island. The whole theory of struggle, as formulated by Darwin, is, therefore, a violent assumption. Men of science now recognize that "egoism and struggle play a very subordinate part in organic development, in comparison with co-operation and social action." What, indeed, but a surrender of the paramountcy of struggle for life, is Huxley's celebrated Romanes lecture in which he supplants the cosmic process by the ethical? The French free-thinker, Charles Robin, gave expression to the verdict of exact science when he declared: "Darwinism is a fiction, a poetical accumulation of probabilities without proof, and of attractive explanations without demonstration."

2. The hopeless inadequacy of the struggle for life to account for adaptive structures has been dealt with at considerable length by Professor Morgan in the concluding chapters of the work already mentioned. We cannot here follow him in his study of the various kinds of adaptations, e.g., form and symmetry, mutual adaptation of colonial forms, protective coloration, organs of extreme perfection, tropisms and instincts, etc., in regard to the origin of each of which he is forced to abandon the Darwinian theory. It will suffice to call attention to his conclusions concerning the phenomena of regeneration of organs. By his research in this special field Professor Morgan has won international recognition among men of science. It was while prosecuting his studies in this field that he became impressed with the utter bankruptcy of the theory of natural selection which Darwinians put forward to explain the acquisition by organisms of this most useful power of regeneration. It is not difficult to show that regeneration could not in many cases, and presumably in none, have been acquired through natural selection (p. 379). If an earth worm (allolobophora foctida) be cut in two in the middle, the posterior piece regenerates at its anterior cut end, not a head but a tail. "Not by the widest stretch of the imagination can such a result be accounted for on the selection theory." Quite the reverse case presents itself in certain planarians. If the head of planaria

lugubris is cut off just behind the eyes, there develops at the cut surface of the head-piece another head turned in the opposite direction. "These and other reasons," concludes Professor Morgan (p. 381), "indicate with certainty that regeneration cannot be explained by the theory of natural selection."

The ingenuity of the Darwinian imagination, however, will hardly fail to assign some reason why two heads are more useful than one in the above instance, and thus reconcile the phenomenon with Darwinism. For, according to Professor Morgan "to imagine that a particular organ is useful to its possessor and to account for its origin because of the imagined benefit conferred, is the general procedure of the followers of the Darwinian school." "Personal conviction, mere possibility," writes Quatrefages, "are offered as proofs, or at least as arguments in favor of the theory." "The realms of fancy are boundless," is Blanchard's significant comment on Darwin's explanation of the blindness of the mole. "On this class of speculation," says Bateson in his "Materials for the Study of Variation," referring to Darwinian speculation as to the beneficial or detrimental nature of variations, "on this class of speculation the only limitations are those of the ingenuity of the author." The general form of Darwin's argument, declared the writer of a celebrated article in the North British Review, is as follows: "All these things may have been, therefore my theory is possible; and since my theory is a possible one, all those hypotheses which it requires are rendered probable."

3. We pass now to the question of the possibility of building up a new species by the accumulation of chance individual variations. That species ever originate in this way is denied by the advocates of the evolutionary theory which is now superseding Darwinism. Typical of the new school is the botanist Hugo De Vries of Amsterdam. The "first-steps" in the origin of new species according to De Vries are not fluctuating individual variations, but mutations, i.e., definite and permanent modifications. According to the mutation theory a new species arises from the parent species, not gradually but suddenly. It appears suddenly "without visible preparation and without transitional steps." The wide acceptance with which this theory is meeting must be attributed to the fact that men of science no longer believe in the origin of species by the accumulation of slight fluctuating modifications. To quote the words of De Vries, "Fluctuating variation cannot overstep the limits of the species, even after the most prolonged selection--still less can it lead to the production of new, permanent characters." It has been the wont of

Darwinians to base their speculations on the assumption that "an inconceivably long time" could effect almost anything in the matter of specific transformations. But the evidence which has been amassed during the past forty years leaves no doubt that there is a limit to individual variability which neither time nor skill avail to remove. As M. Blanchard asserts in his work, La vie des etres animes (p. 102), "All investigation and observation make it clear that, while the variability of creatures in a state of nature displays itself in very different degrees, yet, in its most astonishing manifestations, it remains confined within a circle beyond which it cannot pass."

It is interesting to observe how writers of the Darwinian school attempt to explain the origin of articulate language as a gradual development of animal sounds. "It does not," observes Darwin, "appear altogether incredible that some unusually wise ape-like animal should have thought of imitating the growl of a beast of prey, so as to indicate to his fellow monkeys the nature of the expected danger. And this would have been a first step in the formation of a language." But what a tremendous step! An ape-like animal that "thought" of imitating a beast must certainly have been "unusually wise." In bridging the chasm which rational speech interposes between man and the brute creation, the Darwinian is forced to assume that the whole essential modification is included in the first step. Then he conceals the assumption by parcelling out the accidental modification in a supposed series of transitional stages. He endeavors to veil his inability to explain the first step, as Chevalier Bunsen remarked, by the easy but fruitless assumption of an infinite space of time, destined to explain the gradual development of animals into men; as if millions of years could supply the want of an agent necessary for the first movement, for the first step in the line of progress. "How can speech, the expression of thought, develop itself in a year or in millions of years, out of unarticulated sounds which express feelings of pleasure, pain, and appetite? The common-sense of mankind will always shrink from such theories."

4. The hopes and fears of Darwinians have rightly been centered on the history of organic development as outlined in the geological record. It has been pointed out repeatedly by the foremost men of science that if the theory of genetic descent with the accumulation of small variations be the true account of the origin of species, a complete record of the ancestry of any existing species would reveal no distinction of species and genera. Between any two well-defined species, if one be derived from the other, there must be

countless transition forms. But palaeontology fails to support the theory of evolution by minute variations. Darwinism has been shattered on the geologic rocks. "The complete absence of intermediate forms," says Mr. Carruthers, "and the sudden and contemporaneous appearance of highly organized and widely separated groups, deprive the hypothesis of genetic evolution of any countenance from the plant record of these ancient rocks. The whole evidence is against evolution (i.e., by minute modification) and there is none for it." (cf. History of Plant Life and its Bearing on Theory of Evolution, 1898). Similar testimony regarding the animal kingdom is borne by Mr. Mivart in the following carefully worded statement: "The mass of palaeontological evidence is indeed overwhelmingly against minute and gradual modification." "The Darwinian theory," declared Professor Fleischmann of Erlangen, recently, "has not a single fact to confirm it in the realm of nature. It is not the result of scientific research, but purely the product of the imagination."

On one occasion Huxley expressed his conviction that the pedigree of the horse as revealed in the geological record furnished demonstrative evidence for the theory of evolution. The question has been entered into in detail by Professor Fleischmann in his work, Die Descendenstheorie. In this book the Erlangen professor makes great capital out of the "trot-horse" (Paradepferd) of Huxley and Haeckel; and as regards the evolutionary theory, easily claims a verdict of "not proven." In this connection the moderate statement of Professor Morgan is noteworthy: "When he (Fleischmann) says there is no absolute proof that the common plan of structure must be the result of blood relationship, he is not bringing a fatal argument against the theory of descent, for no one but an enthusiast sees anything more in the explanation than a very probable theory that appears to account for the facts. To demand an absolute proof is to ask for more than any reasonable advocate of the descent theory claims for it." (Professor Morgan, as we have already seen, rejects Darwinism, and inclines to the mutation theory of De Vries.) The vast majority of Darwinians must, therefore, be classed as "enthusiasts" who are not "reasonable advocates of the descent theory." For has not Professor Marsh told his readers that "to doubt evolution is to doubt science?" And similar assertions have been so frequently made and reiterated by Darwinians that the claim that Darwinism has become a dogma contains, as Professor Morgan notes, more truth than the adherents of that school find pleasant to hear.

More interesting, however, than Huxley's geological pedigree of the horse is

Haeckel's geological pedigree of man. One who reads Haeckel's Natural History of Creation can hardly escape the impression that the author had actually seen specimens of each of the twenty-one ancestral forms of which his pedigree of man is composed. Such, however, was not the case. Quatrefages, speaking of this wonderful genealogical tree which Haeckel has drawn up with such scientific accuracy of description, observes: "The first thing to remark is that not one of the creatures exhibited in this pedigree has ever been seen, either living or in fossil. Their existence is based entirely upon theory." (Les Emules de Darwin, ii. p. 76). "Man's pedigree as drawn up by Haeckel," says the distinguished savant, Du Bois-Reymond, "is worth about as much as is that of Homer's heroes for critical historians."

In constructing his genealogies Haeckel has frequent recourse to his celebrated "Law of Biogenesis." The "Law of Biogenesis" which is the dignified title Haeckel has given to the discredited recapitulation theory, asserts that the embryological development of the individual (ontogeny), is a brief recapitulation, a summing up, of the stages through which the species passed in the course of its evolution in the geologic past, (phylogeny). Ontogeny is a brief recapitulation of phylogeny. This, says Haeckel, is what the "fundamental Law of Biogenesis" teaches us. (The reader of Haeckel and other Darwinians will frequently find laws put forward to establish facts: whereas other men of science prefer to have facts establish laws). When, therefore, as Quatrefages remarks, the transition between the types which Haeckel has incorporated into his genealogical tree, appears too abrupt, he often betakes himself to ontogeny and describes the embryo in the corresponding interval of development. This description he inserts in his genealogical mosaic, by virtue of the "Law of Biogenesis."

Many theories have been constructed to explain the phenomena of embryological development. Of these the simplest and least mystical is that of His in the great classic work on embryology, "Unsere Koerperform." His tells us: "In the entire series of forms which a developing organism runs through, each form is the necessary antecedent step of the following. If the embryo is to reach the complicated end-form, it must pass, step by step, through the simpler ones. Each step of the series is the physiological consequence of the preceding stage, and the necessary condition for the following." But whatever theory be accepted by men of science, it is certainly not that proposed by Haeckel. Carl Vogt after giving Haeckel's statement of the "Law of Biogenesis" wrote: "This law which I long held as well-founded, is absolutely and radically false." Even Oskar Hertwig, perhaps the best known of Haeckel's former pupils, finds it necessary to change Haeckel's expression of the biogenetic law so that "a contradiction contained in it may be removed." Professor Morgan, finally, rejects Haeckel's boasted "Law of Biogenesis" as "in principle, false." And he furthermore seems to imply that Fleischmann merits the reproach of men of science, for wasting his time in confuting "the antiquated and generally exaggerated views of writers like Haeckel."

"Antiquated and generally exaggerated views." Such is the comment of science on Haeckel's boast that Darwin's pre-eminent service to science consisted in pointing out how purposive adaptations may be produced by natural selection without the direction of mind just as easily as they may be produced by artificial selection and human design. And yet the latest and least worthy production from the pen of this Darwinian philosopher, The Riddle of the Universe, is being scattered broad-cast by the anti-Christian press, in the name and guise of popular science. It is therein that the evil consists. For the discerning reader sees in the book itself, its own best refutation. The pretensions of Haeckel's "consistent and monistic theory of the eternal cosmogenetic process" are best met by pointing to the fact that its most highly accredited and notorious representative has given to the world in exposition and defense of pure Darwinian philosophy, a work, which, for boldness of assertion, meagerness of proof, inconsequence of argument, inconsistency in fundamental principles and disregard for facts which tell against the author's theory, has certainly no equal in contemporary literature. In the apt and expressive phrase of Professor Paulsen, the book "fairly drips with superficiality" (von Seichtigkeit triefen). If the man of science is to be justified, as Huxley suggested, not by faith but by verification, Haeckel and his docile Darwinian disciples have good reason to tremble for their scientific salvation.

EDWIN V. O'HARA.

St. Paul, Minn.

INTRODUCTION.

During the last few years I have published under this title short articles dealing with the present status of Darwinism. In view of the kind reception which has been accorded to these articles by the reading public I have thought it well to bring them together in pamphlet form. Indeed, the Darwinian movement and its present status are eminently deserving of consideration, especially on the part of those before whom Darwinism has hitherto always been held up triumphantly as a scientific disproof of the very foundations of the Christian faith.

By way of introduction and explanation some general preliminary remarks may not be amiss here. Previous to twenty or thirty years ago, it was justifiable to identify Darwinism with the doctrine of Descent, for at that time Darwinism was the only doctrine of Descent which could claim any general recognition. Consequently, one who was an adherent of the doctrine of Descent was also a Darwinian. Those to whom this did not apply were so few as to be easily counted. The dispute then hinged primarily on Darwinism; hence, for those who did not admit the truth of that theory, the doctrine of Descent was for the most part also a myth.

I say, for the most part; for there were already even at that time a few clear-sighted naturalists (Wigand, Naegeli, Koelliker and others) who saw plainly the residue of truth that would result from the discussion. But to the overwhelming majority, the alternatives seemed to be: Either Darwinism or no evolution at all. Today, however, the state of things is considerably altered. The doctrine of Descent is clearly and definitely distinguished from Darwinism at least by the majority of naturalists. It is therefore of the utmost importance that this luminous distinction should likewise become recognized in lay circles.

My object in these pages is to show that Darwinism will soon be a thing of the past, a matter of history; that we even now stand at its death-bed, while its friends are solicitous only to secure for it a decent burial.

Out of the chaos of controversy which has obtained during the last four decades there has emerged an element of truth--for there lurks a germ of truth in most errors--which has gained almost universal recognition among contemporary men of science, namely, the doctrine of Descent. The fact that living organisms form an ascending series from the less perfect to the more perfect; the further fact that they also form a series according as they display more or less homology of structure and are formed according to similar types;

and, lastly, that the fossil remains of organisms found in the various strata of the earth's surface likewise represent an ascending series from the simple to the more complex--these three facts suggested to naturalists the thought that living organisms were not always as we find them to-day, but that the more perfect had developed from simpler forms through a series of modifications. These thoughts were at first advanced with some hesitation, and were confined to narrow circles. They received, however, material support when, during the fourth decade of the 19th century the splendid discovery was made (by K. E. von Baer) that every organism is slowly developed from a germ, and in the process of development passes through temporary lower stages to a permanent higher one. Even at that time many naturalists believed in a corresponding development of the whole series of organisms, without of course being able to form a clear conception of the process. Such was the state of affairs when Darwin in the year 1859 published his principal work, The Origin of Species by Means of Natural Selection. In this work for the first time an exhaustive attempt was made to sketch a clear and completely detailed picture of the process of development.

Darwin started with the fact that breeders of animals and growers of plants, having at their disposal a large number of varieties, always diverging somewhat from each other, choose individuals possessing characteristics which they desired to strengthen, and use only these for procreation. In this manner the desired characteristic is gradually made more prominent, and the breeder appears to have obtained a new species. Similar conditions are supposed to prevail in Nature, only that there is lacking the selecting hand of the breeder. Here the so-called principle of Natural Selection holds automatic sway by means of the Struggle for Existence. All the various forms of life are warring for the means of subsistence, each striving to obtain for itself the best nourishment, etc. In this struggle those organisms will be victorious which possess the most favorable characteristics; all others must succumb. Hence those only will survive which are best adapted to their environment. But between those which survive, the struggle begins anew, and when the favoring peculiarities become more pronounced in some, (by chance, of course) these in turn win out. Thus Nature gradually improves her various breeds through the continued action of a self-regulating mechanism. Such are the main features of Darwinism, its real kernel, about which of course,--and this is a proof of its insufficiency,--from the very beginning a number of auxiliary hypotheses attached themselves.

Darwin's theory sounds so clear and simple, and seems at first blush so luminous that it is no wonder if many careful naturalists regarded it as an incontrovertible truth. The warning voice of the more prudent men of science was silenced by the loud enthusiasm of the younger generation over the solution of the greatest of the world-problems: the genesis of living beings had been brought to light, and--a thing which admitted of no doubt--man as well as the brute creation was a product of purely natural evolution. The doctrine which materialism had already proclaimed with prophetic insight, had at length been irrefragably established on a scientific basis: God, Soul and Immortality were contemptuously relegated to the domain of nursery tales. What further use was there for a God when, in addition to the Kant-Laplacian theory of the origin of the planetary system, it had been discovered that living organisms had likewise evolved spontaneously? How could man who had sprung from the irrational brute possess a soul? And thus, finally, disappeared the third delusion, the hope of immortality. For with death the functions of the body simply cease, as also do those of the brain, which people had foolishly believed to be something more than an aggregation of atoms. The body dissolves into its constituent elements and serves in its turn to build up other organisms: but as a human body it all turns to dust nor 'leaves a wrack behind'. Thus Darwinism was made the basis first for a materialistic, and then for a monistic, view of the world, and hence came to be rigorously opposed to every form of Theism. But since, at that time, Darwinism was the only theory of evolution recognized by the world of science, the opposition of the Christian world was directed not specifically against Darwinism, but against the theory of evolution as such. The wheat was rooted up with the tares.

I will not discuss here which of the two views concerning creation; the origin of the world in one moment of time, or a gradual evolution of the world and its potentialities, is the more worthy of the creative power of God. Manifestly the greatness and magnificence of creation will in no way be compromised by the concept of evolution. This, of course, is simply my opinion. Any further statement would be out of place here.

But what is the Darwinian position?

It is merely a special form of the evolutionary theory, one of the various attempts to explain how the process of development actually took place.

Darwinism as understood in the following chapters possesses the following characteristic traits:

(1) Evolution began and continues without the aid or intervention of a Creator.

(2) In the production of Variations there is no definite law; Chance reigns supreme.

(3) There is no indication of purpose or finality to be detected anywhere in the evolutionary process.

(4) The working factor in evolution is Egoism, a war of each against his fellows: this is the predominating principle which manifests itself in Nature.

(5) In this struggle the strongest, fleetest and most cunning will always prevail, (the Darwinian term "fittest" has been the innocent source of a great deal of error).

(6) Man, whether you regard his body or his mind, is nothing but a highly developed animal.

A careful examination of Darwinism shows that these are the necessary presuppositions, or, if you will, the inevitable consequences of that theory. To accept that theory is to repudiate the Christian view of the world. The truth of the above propositions is utterly incompatible, not only with any religious views, but with our civil and social principles as well.

The most patent facts of man's moral life, however, cannot be explained on any such hypothesis, and the logic of events has already shown that Darwinism could never have won general acceptance but for the incautious enthusiasm of youth which intoxicated the minds of the rising generation of naturalists and incapacitated them for the exercise of sober judgment. To show that there is among contemporary men of science a healthy reaction against Darwinism is the object of this treatise.

The reader may now ask, What, then, is your idea of evolution? It certainly is easier to criticise than to do constructive work. An honest study of nature, however, inevitably leads us to the conclusion that the final solution of the problem is still far distant. Many a stone has already been quarried for the future edifice of evolution by unwearied research during the last four decades. But in opposition to Darwinism it may, at the present time, be confidently asserted that any future doctrine of evolution will have to be constructed on the following basic principles:

(1) All evolution is characterized by finality; it proceeds according to a definite plan, and tends to a definite end.

(2) Chance and disorder find no place in Nature; every stage of the evolutionary process is the result of law-controlled factors.

(3) Egoism and struggle among living organisms are of very subordinate importance in comparison with co-operation and social action.

(4) The soul of man is an independent substance, and entirely unintelligible as a mere higher stage of development of animal instinct.

A theory of evolution, however, resting on these principles cannot dispense with a Creator and Conserver of the world and of life.

CHAPTER I.

"It was a happy day that people threw off the straight-jacket of logic and the burdensome fetters of strict method, and mounting the light-caparisoned steed of philosophic science, soared into the empyrean, high above the laborious path of ordinary mortals. One may not take offense if even the most sedate citizen, for the sake of a change, occasionally kicks over the traces, provided only that he returns in due time to his wonted course. And now in the domain of Biology, one is led to think that the time has at length arrived for putting an end to mad masquerade pranks and for returning without reserve to serious and sober work, to find satisfaction therein." With these words did the illustrious Wigand, twenty-five years ago, conclude the preface to the third volume of his large classical work against Darwinism. True, he did not at that time believe that the mad campaign of Darwinism had already ended to its own detriment, but he always predicted with the greatest confidence that the struggle would soon terminate in victory for the anti-Darwinian camp. When Wigand closed his eyes in death in 1896, he was able to bear with him the consciousness that the era of Darwinism was approaching its end, and that he had been in the right.

Today, at the dawn of the new century, nothing is more certain than that Darwinism has lost its prestige among men of science. It has seen its day and will soon be reckoned a thing of the past. A few decades hence when people will look back upon the history of the doctrine of Descent, they will confess that the years between 1860 and 1880 were in many respects a time of carnival; and the enthusiasm which at that time took possession of the devotees of natural science will appear to them as the excitement attending some mad revel.

A justification of our hope that Wigand's warning prediction will finally be fulfilled is to be found in the fact that to-day the younger generation of naturalists is departing more and more from Darwinism. It is a fact worthy of special mention that the opposition to Darwinism to-day comes chiefly from the ranks of the zoologists, whereas thirty years ago large numbers of zoologists from Jena associated themselves with the Darwinian school, hoping to find there a full and satisfactory solution for the profoundest enigmas of natural science.

The cause of this reaction is not far to seek. There was at the time a whole group of enthusiastic Darwinians among the university professors, Haeckel leading the van, who clung to that theory so tenaciously and were so zealous in propagating it, that for a while it seemed impossible for a young naturalist to be anything but a Darwinian. Then the inevitable reaction gradually set in. Darwin himself died, the Darwinians of the sixties and seventies lost their pristine ardor, and many even went beyond Darwin. Above all, calm reflection took the place of excited enthusiasm. As a result it has become more and more apparent that the past forty years have brought to light nothing new that is of any value to the cause of Darwinism. This significant fact has aroused doubts as to whether after all Darwinism can really give a satisfactory explanation of the genesis of organic forms.

The rising generation is now discovering what discerning scholars had already recognized and stated a quarter of a century ago. They are also returning to a study of the older opponents of Darwinism, especially of

Wigand. It is only now, many years after his death, that a tribute has been paid to this distinguished savant which unfortunately was grudgingly withheld during his life. One day recently there was laid before his monument in the Botanical Garden of Marburg a laurel-wreath with the inscription: "To the great naturalist, philosopher and man." It came from a young zoologist at Vienna who had thoroughly mastered Wigand's great anti-Darwinian work, an intelligent investigator who had set to work in the spirit of Wigand. Another talented zoologist, Hans Driesch, dedicates to the memory of Wigand two books in rapid succession and reprehends the contemporaries of that master of science for ignoring him. O. Hammann abandons Darwinism for an internal principle of development. W. Haacke openly disavows Darwinism; and even at the convention of naturalists in 1897, L. Wilser was allowed to assert without contradiction that, "anyone who has committed himself to Darwinism can no longer be ranked as a naturalist."

These are all signs which clearly indicate a radical revolution, and they are all the more significant since it is the younger generation, which will soon take the lead, that thinks and speaks in this manner. But it is none the less noteworthy that the younger naturalists are not alone in this movement. Many of the older men of science are swelling the current. We shall recall here only the greatest of those whom we might mention in this connection.

Julius von Sachs, the most gifted and brilliant botanist of the last century, who unfortunately is no longer among us, was in the sixties an outspoken Darwinian, as is evident especially from his History of Botany and from the first edition of his Handbook of Botany. Soon, however, Sachs began to incline toward the position assumed by Naegeli; and as early as 1877, Wigand, in the third volume of his great work, expressed the hope that Sachs would withdraw still further from Darwinism. As years went by, Sachs drifted more and more from his earlier position, and Wigand was of opinion that to himself should be ascribed the credit of bringing about the change. During his last years Sachs had become bitterly opposed to Darwinism, and in his masterly "Physiological Notes" he took a firm stand on the "internal factors of evolution."

During recent years I had the pleasure of occasional correspondence with Sachs. On the 16th of September, 1896, he wrote me: For more than twenty years I have recognized that if we are to build up a strictly scientific theory of organic structural processes, we must separate the doctrine of Descent from Darwinism. It was with this intention that he worked during the last years of his life and it is to be hoped that his school will continue his researches with this aim in view.

The tendency among naturalists to return to Wigand is well exemplified in an article contributed to the "Preussischen Jahrbuecher" for January, 1897, by Dr. Karl Camillo Schneider, assistant at the zoological Institute of the University of Vienna. This article which is entitled The Origin of Species, pursues Wigand's train of thought throughout, and whole sentences and even paragraphs are taken verbatim from his main work. This, at all events, is a very instructive indication of the present tendency which deserves prominence: and its significance becomes more evident when we recall how the work of Wigand was received by the non-christian press a quarter of a century ago. It was either ridiculed or ignored. The two methods of treatment were applied to his writings which are always readily employed when the critic has nothing pertinent to say. It is interesting to note that Darwin himself employed this method. Wigand once told me that he had sent Darwin a copy of his work and had addressed a letter to him at the same time merely stating that he had sent the book, making no reference to the line of thought contained in it. Darwin answered immediately in the kindest manner that he had not as yet received the book, but when it arrived he would at once make a careful study of its contents. Darwin did not write to him again, and when a new edition of his works appeared, the work of Wigand, the most comprehensive answer to Darwin ever written, was passed over without even a passing mention. Thus Darwin completely ignored his keenest antagonist.

As has been said, the majority of those who wrote about Wigand ridiculed him: very few regarded him seriously, and even these indulged chiefly in personal recriminations. Thus matters stood twenty-five years ago. Wigand's prediction passed unheeded. That a periodical not having a specifically Christian circle of readers should now publish a condemnation of Darwinism entirely in accordance with the views of Wigand, is a fact which indicates a notable change of sentiment during the intervening years. I should not be at all astonished if many who sneered at Wigand twenty years ago, now read the article in the Preussischen Jahrbuecher with entire approval. Ill-will towards Wigand has not altogether disappeared even to-day. This is evident from the fact that as yet Dr. Schneider does not venture to defend Wigand publicly, nor to acknowledge him as his principal authority. We must be content, however, if only, the truth will finally prevail.

CHAPTER II.

Striking testimony relative to the present position of Darwinism is borne by the Strasburg zoologist, Dr. Goette, who has won fame by his invaluable labors as an historian of evolutionary theory. In the "Umschau," No. 5, 1898, he discusses the "Present Status of Darwinism," and the conclusions he arrives at, are identical with mine. At the outset Goette indicates the distinction between Darwinism and the doctrine of Descent, and then points out that the distinguishing features of the former consist not so much in the three facts of Heredity, Variation, and Over-production, but rather in Selection, Survival of the Fittest, and also in that mystical theory of heredity--the doctrine of Pangenesis--which is peculiarly Darwinian. Since this theory of Pangenesis has found no adherents, the question may henceforth be restricted to the doctrine of natural selection. This Goette very well observes.

He points, moreover, to the fact that the misgivings that were entertained concerning the doctrine of natural selection on its first appearance, were, on the whole, precisely the same as they are to-day; only with this difference, that formerly they were disregarded by naturalists whose clearness of vision was obscured by excessive enthusiasm; whereas, to-day men have again returned to their sober senses and lend their attention more readily to objections.

Goette recalls the fact that M. Wagner tried to supplement natural selection with his "Law of Migration," and that later on, Romanes and Gulick endeavored to supply the evident deficiencies in Darwin's theory, by invoking other principles; and that even at that time, Askenasy, Braun, and Naegeli--and more recently, the lately deceased Eimer--insisted on the fact of definitely ordered variations, in opposition to the theory of Selection.

Many naturalists recognize the difficulties but do not abandon the theory of Selection, thinking that some supplementary principle would suffice to make it acceptable: many others refuse to decide either for or against Darwinism and maintain towards it an attitude of indifference. The younger investigators, however, are utterly opposed to it. "There can be no doubt that since its first appearance the influence of Darwinism on men's minds has notably diminished, although the theory has not been entirely discarded."--But the very fact that the younger naturalists are hostile to it, makes it evident that Darwinism has a still darker future in store for it: that sooner or later it will come to possess a merely historical interest.

"The present position of Darwinism," says Goette, "is characterized especially by the uncertainty of criticism which is unable to declare definitely in favor of either side." Goette finds the chief cause of this uncertainty in the fact "that men of science (even Darwin himself) have widened the concept of selection as a means of originating new species through the interaction of individuals in the same species, so as to express the mutually antagonistic relations existing between several such species." The latter alone is subject to experimental verification, but it can only cause the isolation of existing forms and is not a species-originating selection--with which alone we are here concerned. This kind of selection can enfeeble the existing flora and fauna, but cannot produce a new species. Selection productive of new species "is not actually demonstrable; it is a purely theoretical invention."

Goette next points out that the investigator is everywhere confronted by definitely-directed variation: a fact which does not harmonize with the theory of selection, nor, consequently with Darwinism. If some scientists have not as yet accepted Eimer's presentation of this doctrine, their action is most probably to be attributed to the fear lest "they should have to accept not merely, variation according to definite laws, but likewise a principle of finality and other causes lying beyond the range of scientific investigation." The rejection of the theory of selection often promotes, as Goette rightly observes, a reactionary tendency towards a priori explanations of phenomena with which we are but slightly acquainted. "There are naturalists who do not discard the theory of selection simply because it seems to furnish a much-desired mechanical explanation of purposive adaptions" (a momentous admission to which we shall have occasion to revert).

Others have broken entirely with selection and the principle of utility and extend the idea of finality to the general capacity of organisms to persist. Thus adaptation becomes a principle which transcends the limits of natural science and pervades the whole domain of life. Goette observes that Darwin spoke of useful, less useful and indifferent organisms, by which he meant those adaptations destined for particular vital functions which tend to make the organs more and more specialized. Since the ability to live is threatened by this specialization it cannot be purposive. This is not wholly true, because the more specialized the individual organ becomes, the more perfect is the whole organism which is composed of these specialized organs. The functions of the individual organ may be restricted, but the power of the entire organism is notably increased, according to the law of the division of labor. Goette therefore has not sufficient grounds for rejecting this expression. He considers that a real and permanent purpose for the individual living forms is out of the question, but that this purpose may be sought for in the development and history of the collective life of nature. Definitely ordered variation, he thinks, a scientific explanation of which is indeed yet forthcoming, will explain adaptation equally as well as does selection. After what has been said this statement of Goette must come as a surprise, for one would think that according to his view definite variation explains adaptations better than selection. Goette sums up his main conclusion in the following words: "The doctrine of Heredity or of Descent, which comes from Lamarck though it was first made widely known by Darwin, has since continually gained a broader and surer foundation. But Darwin's own doctrine regarding the causes and process of Descent which alone can be called Darwinism, has on the other hand doubtlessly waned in influence and prestige."

This is exactly what we also maintain: The establishment of the theory of Descent in general, and the continual retrogression of Darwinism in particular. Wigand was entirely right when he said that Darwinism would not live beyond the century.

We may, however, derive from the discussions of Goette something else that is of the highest importance, namely, an admission in which is to be found the real and fundamental explanation of the conduct of the majority of naturalists who still cling to Darwinism. It does not consist in the fact that they are convinced of the truth of Darwinism but in their "reluctance to give up the mechanical explanation of finality proposed by Darwin," or rather in the fear of being driven to the recognition of theistic principles. With commendable candor Goette attacks this method of keeping up a system notwithstanding its recognized deficiencies. Goette furthermore points out especially that this recognition is more widespread than one might be able to gather from occasional discussions on the subject.

From the account which Goette gives of the present status of Darwinism we may safely conclude that Darwinism had entered upon a period of decay; it is in the third stage of a development through which many a scientific doctrine has already passed.

The four stages of this development are the following:

1. The incipient stage: A new doctrine arises, the older representatives of the science oppose it partly because of keener insight and greater experience, partly also from indolence, not wishing to allow themselves to be drawn out of their accustomed equilibrium; among the younger generation there arises a growing sentiment in favor of the new doctrine.

2. The stage of growth: the new doctrine continually gains greater favor among the young generation, finding vent in bursts of enthusiasm; some of the cautious seniors have passed away, others are carried along by the stream of youthful enthusiasm in spite of better knowledge, and the voices of the thoughtful are no longer heard in the general uproar, exultingly proclaiming that to live is bliss.

3. The period of decay: the joyous enthusiasm has vanished; depression succeeds intoxication. Now that the young men have themselves grown older and become more sober, many things appear in a different light. The doubts already expressed by the old and prudent during the stage of growth are now better appreciated and gradually increase in weight. Many become indifferent, the present younger generation becomes perplexed and discards the theory entirely.

4. The final stage: the last adherents of the "new doctrine" are dead or at least old and have ceased to be influential, they sit upon the ruins of a grandeur that even now belongs to the "good old time." The influential and directing spirits have abandoned this doctrine, once so important and seemingly invincible, for the consideration of living issues and the younger generation regards it as an interesting episode in the history of science.

With reference to Darwinism we are in the third stage which is characterized especially by the indifference of the present middle-aged generation and by growing opposition on the part of the younger coming generation. This very characteristic feature is brought into prominence by the discussion of Goette. If all signs, however, are not deceptive, this third stage, that of decay, is drawing to an end; soon we shall enter the final stage and with that the tragic-comedy of Darwinism will be brought to a close.

If some one were to ask me how according to the count of years, I should determine the extent of the individual stages of Darwinism, this would be my answer:

1. The incipient stage extends from 1859 (the year during which Darwin's principal work, The Origin of Species, appeared) to the end of the sixties.

2. The stage of growth: from that time, for about 20 years, to the end of the eighties.

3. The stage of decay: from that time on to about the year 1900.

4. The final stage: the first decade of the new century.

I am not by choice a prophet, least of all regarding the weather. But I think it may not be doubted that the fine weather, at least, has passed for Darwinism. So having carefully scanned the firmament of science for signs of the weather, I shall for once make a forecast for Darwinism, namely: Increasing cloudiness with heavy precipitations, indications of a violent storm, which threatens to cause the props of the structure to totter, and to sweep it from the scene.

CHAPTER III.

As further witnesses to the passing of Darwinism, two botanists may be cited; the first is Professor Korschinsky who in No. 24, 1899, of the Naturwissenschaftliche Wochenschrift published an article on "Heterogenesis and Evolution," which was to be followed later by a large work on this subject. With precision and emphasis he points to the numerous instances in which there occurs on or in a plant, suddenly and without intervention, a variation which may become hereditary under certain circumstances; thus during the last century a number of varieties of garden plants have been evolved. On the basis of such experiments Korschinsky developed the theory which had been proposed by Koelliker in Wuerzburg thirty years earlier, namely, the theory of "heterogeneous production" or "heterogenesis," as Korschinsky calls it. When one understands that a plant gives rise suddenly and without any intervention to a grain of seed, which produces a different plant, it becomes evident that all Darwinistic speculations about selection and struggle for existence are forthwith absolutely excluded. The effect can proceed only from the internal vital powers inherent in the specified organism acting in connection, perhaps, with the internal conditions of life, which suddenly exert an influence in a new direction.

Korschinsky distinguishes clearly and definitely between the principles of Heterogenesis and Transmutation (gradual transformation through natural selection in the struggle for existence), and in so doing comes to a complete denial of Darwinism.

The other naturalist who has dealt Darwinism a telling blow is the botanist of Graz, Professor Haberlandt.

He published some very interesting observations and experiments in the "Festschrift fuer Schwendener" (Berlin 1899, Borntraeger). They are concerned with a Liane javas of the family of mulberry plants (Conocephalus ovatus.) The free leaves possess under the outer layer, a tissue composed of large, thin-walled, water-storing cells; flat cavities on the upper side, having, furthermore, organs that secrete water, which the botanist calls hydathodes. These are delicate, small, glandular cells over which are the bundles of vascular fibres (leaf-veins) that convey the water to them; over these in the top layer are so-called water-crevices through which the water can force itself to the outside. It is unnecessary to enter upon a closer explanation of the anatomical structure of these peculiar organs. The water which is forced upward by the root-pressure of the plant is naturally conveyed through the vascular fibres into the leaves and at every hydathode the superfluous water oozes out in drops, a phenomenon which one can also very nicely observe e.g. on the "Lady's cloak" (Alchemilla vulgaris) of the German flora. A portion of the night-dew must be attributed to this secretion of water. On the Liane, then, Haberlandt observed a very considerable secretion of water: a full-grown leaf secreted during one night 2.76 g. of water (that is 26 per cent. of its own weight.) Through this peculiarity the water supply within the plant is regulated and the danger avoided that any water should penetrate the surrounding tissue in consequence of strong root-pressure,--which would naturally obstruct the vital function of the entire leaf. Besides it is to be noticed that in this way an abundant flow of water is produced: the plant takes up large quantities of water from the earth, laden with nutritive salts, and the distilled water is almost pure (it contains only 0.045 g. salts), so that the nutritive salts are absorbed by the plant.

From these considerations it necessarily appears that the hydathodes are of great biological importance to the plant.

Haberlandt then "poisoned" the plant, by sprinkling it with a 0.1 per cent sublimate solution of alcohol. The purpose of this experiment was to ascertain whether in the secretion of water there was question of a merely physical process or of a vital process. In the first case the action of the hydathode should continue even after the treatment with the sublimate solution, while in the latter case it should not. As the secretion ceased the obvious conclusion to be deduced from this experiment is that the hydathodes do not act as purely mechanical filtration-apparatuses, as one might have thought, but that there is here evidence of an active vital process in the plant; the unusual term "poisoning" is therefore really justified under present circumstances.

Let me dwell for a moment on this result, for, although it may be somewhat foreign to our present purpose and to the further observations of Haberlandt, it is very significant in itself. The water moves in the plant in closed cells, as the cells of the aqueous gland are entirely closed, but the organic membrane, as every one knows, has the peculiar physical property of allowing water to pass through, the pressure, of course, being applied on the side of least resistance; when therefore the water is forced into the cells by root-pressure, it is easily intelligible that according to purely physical laws it should come to the surface of the leaf on the side of the least resistance, that is, by way of the water-crevices. Even the defenders of "vital force" would not find any reason in this for not considering the phenomenon of distillation in this case a purely physical phenomenon. And still according to Haberlandt's experiments it is not. The sublimate could at most only impede the process of filtration, but should under no circumstances have destroyed it. But it does destroy it, and the hydathode dies. The conclusion certainly follows from this that this process is connected with some vital function. Even if the hydathode is treated with sublimate solution, all the conditions for mechanical filtration still remain: the earth has moisture which can be taken up by the roots so that root-pressure still exists. The water is in all cases conveyed to the hydathodes through the vascular fibres, the cell walls of the hydathodes are still adapted for filtration, and yet they do not filter. Hence some other factor must join itself to the physico-mechanical process of filtration and affect or destroy it, and this factor can be found only in the protoplasm, the vital element of the cells; for we know that the sublimate acts with pernicious effect on it and in such a manner that it destroys its entire power of reaction; it kills it, as we say.

The experiment under discussion has, therefore, great significance for our view of the vital processes in the plant; it proves beyond doubt that these processes are in no way of a purely mechanical nature, but that there is something underlying all this, a hitherto inexplicable something, which we call "life." In all vital activities, physical and chemical processes certainly do occur; they do not, however, take place spontaneously but are made use of by the vital element of the plant to produce an effect that is desirable or necessary for the vital activity of the plant. If the vital element is dead, no matter how favorable the conditions may be for chemical and physical processes, these do not take place and the effect necessary for life is not obtained. It is very remarkable after all that according to the experiment of Haberlandt this peculiar relation should become apparent in a process that is so open to our investigation as the filtration of water through the cell-wall of a plant.

After what has been said I consider this simple experiment of Haberlandt of great significance; for it is a direct proof of the existence of a vital force. One may resist to his heart's content, but without avail; vital force is again finding its way into science. More and more cognizance is being taken of the fact that 60 and 70 years ago people jumped at conclusions very imprudently when they believed that the first artificial preparation of organic matter (urea, by Woehler) had proven the non-existence of a vital force. Since then there has been great rejoicing in the camp of materialists who scoffed at the "ignorant" who would not as yet forsake vital force. "Behold," they said, "in the chemist's retort the same matter is produced chemically that is produced in the body of the animal, without the direction of a hidden vital force, which, if it is not necessary in the one case, neither is it necessary in the other." Any one who had given the matter careful consideration could even at that time have known where the "ignorant" really were. That in both cases chemical processes take place is clear and undisputed, but the materialists forgot entirely that even in the laboratory it was not the mere contact of the elements that produced the urea; a chemist was needed and in this case not any one arbitrarily chosen, but a man of the genius and knowledge of a Woehler to watch over the process, and utilize and partly direct the laws of chemistry in order to obtain the desired result. Hence it was even then absurd to deny vital force as a consequence of that experiment. Since, however, it was well-adapted for materialistic purposes, this denial was proclaimed with the sound of trumpet throughout the land, and repeated again and again with surprising tenacity, with the result that even thoughtful investigators rejected vital force almost universally in the seventies and eighties.

It has always been a problem to me how this could have happened. It can, indeed, be explained only on the supposition that naturalists were adverse to the introduction of anything into nature, that appeared to them mystical and mysterious. Nor is such a procedure at all necessary: vital force is by no means a mysterious, ghostly power that soars above nature, but a force of nature like its other forces, as mysterious and as definite as they are, only that it dominates a specified group of beings, namely, living organisms. It may readily be compared with any other natural phenomenon. For instance, the phenomenon of crystallization has its well determined sphere of activity, viz., the mineral world. It employs definite mathematico-physical laws to obtain a specified result, and even acts differently in different mineral substances in so far as it produces in the one case this, in the other case that form; but still it should be a similarly directed force which has the effect of producing these peculiar forms. Precisely similar is it with vital force. It has its determined sphere of activity, the kingdom of living organisms; it acts according to definite physico-chemical laws in producing a specified result; it acts differently in different living organisms; it is therefore a force of nature as clear yet as mysterious as the force of crystallization or as any other force of nature. Hence one has no cause to complain of its mysteriousness, for all other forces of nature are just as much, or if you will, just as little mysterious as vital force. The only thing to be maintained is this, that living organisms are dominated by a special force with special phenomena and special activities, even as in mineral substances there is a special dominant force which produces special phenomena and exercises special activities.

It is possible to produce crystals in the laboratory, but no one will be so foolish as to maintain that in nature crystals are not formed in consequence of a very definite force inherent in the mineral-substances; nor will any one deny the existence of the force of crystallization because it does not appear in living organisms.

Nor have I ever despaired of a return of the theory of vital force. A change of opinion has really taken place during this decade; at present the voices for a vital force are constantly growing stronger and it will most probably not be very long before it will be again universally recognized, not as something preternatural, of course, but as a force of nature on an equal footing with the other forces of nature, with activities, just as mysterious and just as well-attested as the activities of the other forces of nature.

Haberlandt's experiment, however, had also an indirect consequence that is of far-reaching importance. He observed that within a few days new water-secreting organs of an entirely different structure and of different origin were formed on the leaves that had been sprinkled with sublimate. Over the bundles of vascular fibres, little knots as large as a pin head arose in larger numbers out of a tissue underlying the top layer; out of these the water now oozed every morning. Closer investigation disclosed the fact that these organs develop only on young immature leaves where groups of peculiar, perishable gland-hairs are found; beneath these dead mucous glands the substitute secretive organs originate in the inner tissue. It is of no importance to state in what particular cells they originate.

Suffice it to say that they are colorless capillary tubes originating in various cells; projecting like the hairs of a brush, containing living protoplasm and evanescent chlorophyll. It is also important to note that this new organ is immediately connected with the water-conducting system consisting of bundles of vascular fibres. Haberlandt furthermore indicates especially that these organs when viewed in connection with the process of secretion give evidence of an active vital principle as well as of simple mechanical filtration.

These substitute organs are all indeed well adapted to their purpose and adequately replace the old secretive organs, but they so easily dry out and are so little protected that after a week they become parched and die because wound-cork forms under them. The leaf no longer produces new hydathodes, but on its lower side it produces growths that function as vesicles, by means of which it continues to sustain itself.

Haberlandt furthermore records a phenomenon perhaps analogous to this on the grape-vine, but with this exception the case described by him is unique. In order to pass any further judgment regarding it, we should have to ascertain whether the whole phenomenon is not a case of so-called adaptation; if so, processes should be found in nature, analogous to the poisoning of the hydathodes in this experiment, which result in the destruction of the hydathodes so that in consequence the plant would have gained the power of making good the loss, by means of the substitute organs. Such processes, however, (even through poisoning or through parasites) would be very highly improbable. Equally incredible is the alternative possibility that the new organs would be produced by the plant not as a substitute but as a supplementary apparatus when the old ones would not suffice for secretion in case of very large absorption of water. This also must doubtlessly be rejected, as Haberlandt has observed.

Powers of adaptation should, of course, according to Darwinism, be gradually acquired in the struggle for existence, as in that case they should also have stability; but since this is not possessed by the new organs, the presumption is that they do not possess the character of adaptation. They are therefore new organs that originated after an entirely unnatural and unforeseen interference with the normal vital functions and in consequence of a self-regulating activity of the organism.

What then is there in the whole phenomenon worthy of notice with regard to the theory of Descent?

1. An immediately well adapted new organ has here originated very suddenly without any previous incipient formation, without gradual perfection and without stages of transition.

2. In its formation struggle for existence and natural selection are entirely excluded, neither can find any application whatever even according to the newer exposition of Weismann. Haberlandt himself draws this conclusion.

3. If this phenomenon of a suddenly appearing change can take place in the course of the development of the individual, there can be no obvious reason why it should not take place in the same manner (without natural selection or struggle for existence) in the course of the phylogenetic development.

It is manifestly of the greatest importance that in this case a direct, experimental proof has been given that an organ has originated suddenly and without the aid of Darwinian principles. Haberlandt's article is nothing less than a complete renunciation of Darwinism on the part of Haberlandt, a renunciation which we greet with great satisfaction.

In fact one such observation would really suffice to set aside Darwinism and prove the utter insufficiency of its principles to give explanation of the origin of natural species. On the other hand, this observation plainly proves two things: first, that the above mentioned doctrine of Koelliker, now held by Korschinsky is a move in the right direction for the discovery of the causes of descent; and secondly, that the principal cause of the evolution is not to be sought in environment and blind forces but in the systematically working, internal vital principle in plants and animals. With that, however, an important part of the foundation of the mechanical-materialistic view of the world is demolished.

CHAPTER IV.

Since we have heard the verdict of zoologists and botanists concerning Darwinism, it is but right that we should now listen to a palaeontologist, a representative of the science, which investigates the petrified records of the earth's surface, and strives to collect information regarding the world of life during remote, by-gone ages of the earth. It is evident to every-one that the verdict of this science must be of very special importance in passing on the question of the development of living organisms. Darwin himself recognized this at the outset. He and his followers, however, soon perceived that, while the revelations of palaeontology were on the whole favorable to the doctrine of Descent, in so far as they proved the gradual change of organization, in consecutive strata, from the simple to more complex forms, palaeontology revealed nothing that would sustain the Darwinian theory as to the method of that development. As soon as the Darwinians, and first of all Darwin himself, perceived this, they at once brought forward a very cheap subterfuge. Since Darwinism postulates a very gradual, uninterrupted development of living organisms, there must have been an immense number of transition-forms between any two animal or plant species which to-day, although otherwise related, are separated by characteristic features. Consequently, on the

Darwinian hypothesis, all of these transition-forms must have perished for the singular reason that other better organized forms overcame them in the struggle for existence. If therefore the millions of transition-forms were still missing, and the known petrified forms of older strata of the earth did not reveal them, the Darwinians were able to console themselves until from 20 to 40 years ago, with the assertion that our knowledge was still too deficient, that a more thorough investigation of the earth's surface and especially of out-of-the-way parts would eventually bring to light the supposed transition forms. Such assertion affords very poor consolation, and is anything but scientific. The method of natural science consists in establishing general principles on the basis of the materials actually furnished by experiments and observation and not in excogitating general laws and then consoling oneself with the thought that while our knowledge of nature is as yet extremely imperfect, time will furnish the actual material necessary to substantiate our guesses. But since then many a year has come and gone and Darwinism has caused, and for that alone it deserves credit, a diligent research in every field of natural science, and has promoted among palaeontologists a search for the missing transition-forms. The materials of investigation from the field of palaeontology have also wonderfully increased during these decades. Hence it is worth while now at the dawn of the new century to examine this material with a view to its availableness for the theory of Descent and especially for Darwinism.

Professor Steinmann has recently done so in Freiburg in Breisgau, on the occasion of an address as Rector of the University. What conclusions did he reach?

Steinmann declares it to be the primary task of post-Darwinian palaeontology "to arrange the fossil animal and plant-remains in the order of descent and thus to build up a truly natural, because historically demonstrable, classification of the animal and plant-world." At the outset it is to be noted that for various reasons palaeontology is unable to execute this momentous task in its full extent. The evidence of palaeontology is deficient, if for no other reason than that many animal organisms could not be preserved at all on account of their soft bodies; many animal groups have, nevertheless, received an unusual increase (mollusks, radiata, fish, saurians, vertebrates, and dendroid plants).

As regards the attempt made in the sixties to draw up lines of descent, Steinmann repudiates, without, of course, mentioning names, the family tree constructed by Haeckel and his associates as wholly hypothetical and hence unjustified; he rightly remarks that their method smacks of the closet. He finds fault with them chiefly because they predicated actuality of this imaginary family-tree and fancied that the historical research of the future would have but isolated facts to establish.

In speaking of the palaeontological research of the last few decades, Steinmann says: "In the light of recent research, fossil discoveries have frequently appeared less intelligible and more ambiguous than before, and in those cases in which an attempt has been made to bring the descent-system into agreement with the actual facts, the incongruity between the two has become obvious." Thus, for instance, the well-known archaeopteryx is not, as was maintained, a connecting link between reptile and bird, but a member of a blindly ending side branch. In fact palaeontological research has proven incapable of finding the transitions between different species, clearly determined by the theory. But the overwhelming abundance of matter called for new endeavors to master it. It was then further discovered--Steinmann finds an illustration of this fact in the echinodermata--that the well-known "fundamental law of biogenesis" of Haeckel can be accepted only in a very restricted sense and may even lead to conclusions absolutely false. We desire to remark here that a "fundamental principle" should never mislead; if it does so, it is not a fundamental principle.

It is of importance to know that according to palaeontological investigation, empiric systematizing and phylogenetic classification do not always coincide, as, for instance, in the case of the ammonites. Acording to palaeontological investigation the great systematic categories are only grades of organization. Hence present day systematizing is being more and more discarded, and the said categories--as indeed also the lesser groups of forms--must be of polyphyletic origin, that is, they must have descended from different primitive stocks. It may be asked: What bearing has this principle of multiple origins? For a long time reptiles were the predominating vertebrates; when mammals and birds appeared, numerous, varied and strange saurians inhabited land and sea; but "with the end of the chalk-period most saurians seem to have vanished suddenly from the scene, and soon we behold the mainlands and oceans inhabited by mammals of most diverse kinds." The saurians have become almost extinct and the mammal-tribe suddenly shows a most extraordinary variability and power of development. How is either phenomenon to be explained?

"The disappearance of a group of organisms has been preferably explained since the time of Darwin, by defeat in the struggle with superior competitors. If ever an explanation lacked pertinency, it does so in this case, in which the succumbing group is represented by gigantic and well preserved animal forms, widely distributed and accustomed to the most varied methods of nutrition, whereas the competitor appears in the form of small, harmless marsupials. It would be equivalent to a struggle between the elephant and the mouse."

We acknowledge with pleasure this clear rejection of Darwinism on the part of Steinmann.

Steinmann also rejects the natural extinction of those forms, perhaps from the weakness of old age; whether he is wholly warranted in doing so, seems somewhat doubtful. He tries to explain the phenomenon on the basis of the multiple origin of the mammals; and in fact there is already speculation regarding triple origin, viz: tambreets, marsupials, and the other mammals. Now if the latter also possessed a multiple origin, the problem of the extinction of the saurians would, according to Steinmann solve itself. One would not need to consider the number of extinct forms as large as is now done. However, he does not enter upon any closer consideration of this question. But he points out, for instance, that to-day the shells of mollusks (snails and conchylia) are regarded as structures that were acquired only in the course of time for the sake of protection, the disappearance of which, therefore, implied a disadvantage for the respective organisms. This transition would be something extraordinary--"but if on the contrary, one regards the shells as the necessary products of a special kind of assimilation and of the immoveableness of certain parts of the body, the gradual disappearance might well be considered a process which may take place in various animal-groups with a certain regularity in the course of the phyletic development." The snails devoid of shells, for instance, may be derived with certainty from those possessed of shells; this process has very probably also taken place in different genetic lines.

This view is well worth consideration; it stands in sharp opposition, in fundamental principles, to the Darwinian explanation. This calls for special emphasis here. How should one explain the origin of uncrusted mollusks from crusted ones through the struggle for existence, since in such a contest the latter must have had far greater prospect of survival than the former?

This view together with the principle of multiple origin opens up, according to Steinmann, "the prospect of an altered conception of the process of formation of the organic world." According to the new conception, the many extinct forms of antiquity are not, as Darwin supposed, "unsuccessful attempts and continued aberrations of nature"--how this reminds one of that old, naive, much-ridiculed idea that fossils were models that God had discarded as unserviceable--but would gain new life and assume hitherto unsuspected relationship to the present organic creation.

"Science, which seeks after operative causes, at the beginning of the century regarded creation as a multiplicity of phenomena without any causal connection as to their origin. Darwin taught as a fundamental principle the unity and the causal inter-relation of creation, but was not entirely able to save this hypothesis from a violent and sudden death. In the future sketch creation will appear as wholly restricted in itself and lasting, the causes of its limitation lie, up to the time of the intervention of men, solely in the balanced motion of the planet which it peoples."

At the close of his address Steinmann points out that behind the problem of the manner of development, there stands "the unsolved question regarding its operative causes." "Regarding this point," he continues, "opinions have perhaps never been so divergent as they are to-day. The times have passed when the Darwinian explanations were regarded with naive confidence as the alpha and omega of the doctrine of Descent. Not only are the adherents of Darwinian ideas divided among themselves, but the theory of Lamarck, somewhat altered, favored by the results of historical investigation, appears more striking and now seems more in harmony with facts than formerly. What is considered by one as the ruling factor in the evolution of organisms is regarded by another as a "quantite negligeable" or even as the greatest mistake of the century. In this discord of opinions the principle of Descent alone forms the stable pole."

Thus Steinmann, and we can but applaud his conclusions with undisguised pleasure, for they tend throughout in the direction of our anti-Darwinian view, and deal Darwinism another fatal blow. It is also worthy of special note that this time the blow is dealt from the side of palaeontology; for, even if now and again we dissent from Steinmann, in this we fully agree with him that the historical method of considering the evidences of bygone periods of creation is at the very least quite as important for passing correct judgment regarding descent, as is the investigation of contemporary living organisms. Indeed, family-trees were constructed without regard for palaeontology, almost exclusively from an examination of present conditions, and sometimes the author did not even shrink from falsification. This procedure has been bitterly revenged and will take further revenge unless at length a definite end be put to the family-tree nuisance and the respective books instead of being published anew, be relegated to the lumber-room of science, there to turn yellow amid dust and cobwebs--the curious evidence of gross folly. But only have patience, even that time will come.

The conclusions of Steinmann, that are most important for us, may be summarized as follows:

1. The family and transition forms demanded from palaeontology by Darwinism for its family-trees, constructed not empirically but a priori, are nowhere to be found among the abundant materials which palaeontological investigation has already produced.

2. The results of the investigation do not correspond with the family groups drawn up according to the so-called "biogenetic principle," which principle has in fact led men of science into false paths.

3. At best, the biogenetic principle has a limited validity, (we add that later it will undoubtedly follow Darwinism and its family trees into the lumber-room).

4. The results of palaeontology, in so far, for instance, as they testify to the sudden disappearance of the saurians and the advent of mammals, everywhere contradict the Darwinian principle of the survival of the fittest in the struggle for existence.

5. "The time has long passed when the Darwinian explanations were regarded with naive confidence as the alpha and omega of the doctrine of Descent."

6. Only the principle of Descent is universally recognized; the "how" of it, its causes, are to-day entirely a matter of dispute.

CHAPTER V.

The strongest evidence of the decay of Darwinism is to be found in the fact that, since Darwin first enunciated his theory, many and diverse attempts have been made to explain the origin of species on other principles. Names of men, like M. Wagner, Naegeli, Wigand, Koelliker, and Kerner mark these attempts; but of these investigators Naegeli alone proposed a well-developed hypothesis. Finally, however, Eimer, professor of zoology in Tuebingen came forward with a detailed theory of Descent. As early as 1888 he published a comprehensive work dealing with it, under the title: "The Origin of Species by Means of the Transmission of Acquired Characters According to the Laws of Organic Growth." As the title itself indicates, a very marked divergence was even at that time manifesting itself between Eimer and his former teacher and friend, the great defender of Darwinism in Germany, Aug. Weismann, professor of zoology in Freiburg in Breisgau. For, while the latter vigorously attacks the transmission of acquired characters, Eimer's whole theory is founded on this very transmission. Observations regarding the coloring of animals, in fact, form the basis of Eimer's theory.

Eimer attributes the origin of species to "organic growth" by which he means not merely increase in size, but also change of form, etc. This growth does not proceed blindly or aimlessly, but proceeds on rigidly determined lines, which depend upon the structure and constitution of the particular organism. External influences, however, also affect it. Eimer specially emphasizes four points in this connection: 1. This rigidly determined development of a character exhibits well defined, regular stages, and the evolution of each individual repeats the whole series of transformations (the Mueller-Haeckel "biogenetic-law.") 2. New characters are first acquired by strong adult males (the law of male dominance). 3. New characters appear on definite parts of the body, spreading especially from the rear to the front, (the law of undulation). 4. Varieties are stages in the process of development, through which all the individuals of the respective species must pass.

These points indicate how important for Eimer is the transmission of those characters which the parents themselves have acquired in the course of their own development. He conceives that this transmission takes place when the causative influences exert themselves permanently on many succeeding generations. Eimer thinks that in this way the constitution of the respective species is gradually transformed. Besides the effect of external influences (which may vary according to the climate, etc.: Geoffroy St. Hilaire), Eimer mentions as important and active factors in this development, (1). The use and disuse of organs (Lamarck); (2). The struggle for existence (Darwin); (3). The correlation of organs, that is, the inner relation of organs in consequence of which a change in one organ may occasion a sudden change in another organ; (4). Cross fertilization and hybridism.

It is clear that with reference to the factors of evolution Eimer is, and perhaps not unreasonably, an eclectic, whose aim is to do justice to the predecessors of Darwin as well as to Darwin himself. His antagonism to Darwin and Weismann in this work is still quite moderate, although even here it appears with sufficient clearness that selection and the struggle for existence, the two principles peculiarly characteristic of Darwinism, do not give rise to new species, but can at best only separate and differentiate species already existing.

The second part of Eimer's work dealing with the origin of species, which appeared after an interval of ten years, bears the title: "Orthogenesis of Butterflies." The Origin of Species, II. Part (2 tables and 235 illustrations in the text). Leipzig, 1897. In this book substantially the same thoughts occupy the mind of the author as in the former volume, but in many respects they are more mature, and conspicuously more definite and precise. The most salient features are the following:

1. Eimer establishes his theory by means of very minute observations on a definite species of animals, viz., butterflies.

2. He attributes evolution almost exclusively to development along definitely determined lines.

3. He proves the utter untenableness of Darwinian principles and repudiates them unqualifiedly.

4. In a very distinct and severe manner he gives expression to his opposition to his former friend Weismann.

5. He attacks with telling effect the fantastic Darwinian "Mimicry."

In his "General Introduction" Eimer first treats of Orthogenesis in opposition to the Darwinian theory of selection. The very first sentence gives evidence of this antagonism: "According to my investigation, organic growth (Organophysis), which is rendered dependent on the plasm by permanent external influences, climate and nourishment, and the expression of which is found in development along definitely determined lines, (Orthogenesis), is the principal cause of transformation, its occasional interruption and its temporary cessation and is likewise the principal cause of the division of the series of organisms into species."

Lamarck's theory of the use and disuse of organs and Darwin's hypothesis of natural selection are consequently pushed into the background. Here also Eimer at once places himself at variance with Naegeli who had enunciated a similar theory. Naegeli took as a starting point an inherent tendency in every being to perfect itself, thus presupposing an "inner principle of development," and making light of external influences as transforming causes. Eimer flatly contradicts this view. We shall revert to this point in our criticism of his theory. In opposition to the theory of selection, Eimer lays special stress on the fact that its underlying assumption, viz., fortuitous, indefinite variation in many different directions, is entirely devoid of foundation in fact, and that selection, in order to be effective, postulates the previous existence of the required useful characters, whereas the very point at issue is to explain how these characters have originated. Since, therefore, according to Eimer's investigations, there are everywhere to be found only a few, definitely determined lines of variation, selection is incapable of exercising any choice. The development, furthermore, proceeds without regard for utility, since, for instance, the features that characterize a species of plants are out of all reference to utility. "Even if nothing exists that is essentially detrimental, nevertheless very much does exist that bears no reference whatever to immediate good, and was therefore never affected by selection."

Further on, Eimer expresses still more clearly the opposition of his theory to that of Darwin, and in so doing he attacks vigorously the omnipotence of selection, so unreasonably proclaimed by the followers of Darwin. Eimer's theory, consequently, asserts that: "The essential cause of transmutation is organic growth, a definite variation, which, during long periods of time proceeds unswervingly and without reference to utility, in but few directions and is conditioned by the action of external influences, of climate and nourishment." In consequence of an interruption of orthogenesis a stoppage ensues in certain stages of the development, and this stoppage is the great cause of the arrangement of forms in different species. Of vital importance also "is development through different stages (Hetero-epistase), which results in the arrested development of certain characters in an organism, while others progress and still others become retrogressive. As a rule use and disuse are of great efficacy in this regard, and conjointly with these compensation and correlation." Occasionally also irregular development sets in, which proceeds by leaps.

Of course, Eimer could not but in his turn burn incense before Darwin by declaring that he would not dare to cross swords with such a man, while in reality he repudiates all of Darwin's fundamental tenets.

It may be well to state here in addition a few important supplementary considerations: "Development can everywhere proceed in only a limited number of directions because the constitution, the material composition of the body, conditions these directions and prevents variation in all directions." This is an important statement because Eimer clearly expresses therein the difference between his own theory and that of Naegeli. He makes the direction of development dependent on the material composition of the body, whereas Naegeli considers it dependent upon an internal tendency of every being to perfect itself, hence upon a power inherent in the body. Eimer's view therefore tends towards a mechanical explanation, while Naegeli postulates a vital energy. The "internal causes" according to Eimer find their explanation in the material composition of the body. Since the growth of the individual organism depends on this composition and on the external influences, Eimer compares family-development with it and designates the latter as "organic growth." In opposition to Naegeli he maintains that this "organic growth" does not always aim at perfection but often tends to simplification and retrogression.

The following, then, according to Eimer, are the directive principles of variation: (1). The general law of coloration (stripes running lengthwise change into spots, stripes running crosswise change to a uniform color). (2). The law of definitely directed local change (new colors spread from the rear to the front and from above downward or vice versa, old colors disappear in the same directions.) (3). The law of male predominance (males are as a rule one step in advance of the females in development). Female predominance is an exception. (4). The law of age-predominance (new characters appear at a well-advanced age, and at the time of greatest strength). (5). The law of wave-like development (during the course of the formation of the individual organism a series of changes proceed in a definite direction over the body of the animals). (6). The law of independent uniformity of development (the same course of development is pursued in non-related forms and results in similar forms). (7). The law of development through different stages (different characteristics of the same being may develop to a different degree and in different directions). (8). The law of unilateral development (the progeny does not present a complete combination of the characters of the parents but manifests a preponderance of the characteristics of either parent). (9). The law of the reversal of development (the direction of development may reverse and tend towards the starting point). (10). The law of the cessation of development (a protracted cessation of development frequently ensues in one or the other stage).

The origin (perhaps rather the distinction) of species is accounted for principally by the last named law, by means of which Eimer also explains the so-called atavism or reversion. To this law are joined other factors, e.g., development proceeding in leaps, as demonstrated by Koelliker and Heer; local separation (through migration; prevention of fertilization, e.g., the impossibility of cross-fertilization between certain individual organisms) which Romanes had already opposed to natural selection, and crossing.

The second main division of the book is taken up with a very searching and detailed criticism of Weismann. This criticism seems to me entirely warranted; because not only the latter's unintelligible position with regard to natural selection (the repudiation of which he seems to regard as synonymous "with cessation of all investigation into the causal nexus of phenomena in the domain of life") but likewise his fanciful theory of heredity, utterly devoid as it is of any support from actual observation, bespeak an utter lack of qualities essential to a naturalist; and the manner in which he ignores his former pupil and his labors, because they proved embarrassing to him, is entirely unworthy of a man of science.

Eimer devotes special attention to "mimicry"; and indeed he was forced to be very solicitous to dispel this fanciful conception of Darwinism which radically contradicted his own views. Moreover, the untenableness of the mimicry hypothesis must have revealed itself very clearly to him in the course of his investigations regarding the coloring of butterflies. Mimicry, as our readers are well aware, consists in this, that living beings imitate other organisms or even inanimate objects; Darwinism maintains that this is done for the sake of protection against enemies. This phenomenon is said to have been produced by selection. Those animals that possessed, for instance, some similarity to a leaf, in consequence escaped their enemies more easily than others and survived, while those that had no leaf-like appearance succumbed; when this process had been repeated a few times, many animals (butterflies) gradually developed that marvelous leaf-like appearance, which frequently deceives the most practiced eye.

It appears so simple and natural that one need not wonder that this peculiar phenomenon gained many an adherent for Darwinism. But, of course, it is directly opposed to the views of Eimer; and it is for this reason that he endeavors so assiduously to disprove the error of Darwinism in this regard. As the underlying color design of the butterfly Eimer designates eleven longitudinal designs; and the examination of the leaf-like forms leads him to the conclusion, that their appearance always depends on "the unaltered condition or the greater prominence of certain parts of this fundamental design." There is to be observed a shifting of the third band, so that in conjunction with the fourth, which is curved, it forms the mid-rib of the leaf. Eimer finds the cause of this phenomenon in the alteration of the form. The leaf-like form results from an acumination and elongation of the wings, which in turn results from a marked elongation of the rim of the fore-wing. And this again is produced by the proportionately greater growth of one part of the wing-section than of the others.

With reference to the reason of this growth it is of importance to note that experiments, consisting in the application of artificial heat to the chrysales of the swallow-tail and sailor-butterfly, demonstrated that by this means "the fore-wing is drawn out more toward the outer wing-vein, and the rim of the fore-wing becomes more elongated and curved." It is observed, however, that the natural heat-forms of the same genera and species, namely, the summer-forms and those which live in the warm southern climate, exhibit, for instance, in the case of butterflies akin to the sailor, the same features, the elongation and more marked curvature of the fore-rim of the fore-wings and the consequent more extended form, that are produced by the action of artificial heat. Manifestly this is a matter of vital importance for the solution of the question: heat, whether artificial or natural, produces a difference in growth, which results in a change of form and coloring. There is consequently no room for natural selection or the struggle for existence.

The leaf-like form is generally associated with the dark, faded colors of dry leaves, and when this similarity disappears even bright colors appear on the fore-wings. In many cases the resemblance to leaves is very imperfect; different forms of the same species live side by side and among them are to be found those, the resemblance of which to leaves is extremely slight. All these facts, and especially the frequently recurring retrogression of the leaf-like appearance, justify serious doubt regarding the Darwinian assumption, that adaptation was a necessity for the forest-butterflies on account of the protection which it provided.

An eye witness furthermore declares that the butterflies that resemble leaves most closely do not always alight on withered leaves, on which they would be almost invisible, but frequently rest on a green background, against which they show off very clearly, and therefore could not long escape the keen eye of birds. Besides, these butterflies are but seldom pursued by the birds, of which there is question here, and hence are in no need of protection.

The longer Eimer devoted his attention to the origin of this resemblance the more "the poetic picture of the imitated leaf" vanished out of sight, and he became convinced that it involved the necessary expression of the lines of development, which the respective beings were bound to follow, and that there was no question of imitation.

Apart from the resemblance to leaves, by reason of regular changes of color, design, and wing-structure, numerous non-related butterflies often develop such wonderful similarities--which are not, as hitherto supposed, imitations or disguises produced by selection, but are either the outcome of an entirely independent uniformity of development or, at least, of its consequence--that it must be admitted that external similarity may arise by different means and in various ways. These relations of similarity are of such frequent recurrence because of the limited number of directions of development in which changes or color and design in butterflies may tend. Eimer finds the reason of this small number of directions, in which development may proceed, in the fact "that the elementary external influences of climate and nourishment on the constitution of the organism are everywhere the cause of the transformations."

Another important point is the difference of sex. If the butterflies are of different sex, the males as a rule exhibit a more developed stage of design and color than the females. These frequently present on the upper side the stage of coloration, which the males present on the lower side, while the upper side of the males is one stage in advance. It is of special significance that the characters of the more advanced sex frequently correspond to those of a related, superior species, and occasionally to those of widely separated species. Eimer endeavors to explain male predominance "by a more delicate and more developed, i.e., more complex, chemico-physical organization of the male organism." Even this development tends toward simplification, the origin of dull-black colors.

This most interesting question brings Eimer into conflict with another Darwinian principle, the so-called principle of "sexual election," according to which the more striking characteristics of the male sex become strengthened for the reason that females invariably give the preference to the males endowed with them, over those that are less "attractive." These exceedingly romantic ideas have been often and deservedly repudiated, e.g., even by Wallace only a short time after their first appearance. Eimer really does them too much honor when he again undertakes, even with a certain amount of respect, a thorough refutation of them, "as in every regard unfounded." It is of primary importance to note here, that in the case of dimorphism of the sexes abrupt modifications occur in connection with unilateral heredity. "It is impossible for sexual selection to produce a change of design and color, which results in the sudden kaleidoscopic formation of wholly different designs, as we find actually taking place through the action of artificial heat and cold and other factors in nature."

This brings us to a brief consideration of the answer, which Eimer proposes to give to the question of the real causes of the formation of species among butterflies. A precise and clear statement of this important part of Eimer's theory of Descent, is contained in the following extracts: "The transformation of organisms is primarily conditioned by the action of immediate external influences on the organisms. The same causes, which produce individual growth, especially climate and nourishment, also produce the organic growth of organisms, that is, transmutation, which is but a continuation in the progeny of individual growth, through the transmission of the characteristics acquired during the lifetime of the individual."

Hence, transmutation is simply a physiological process, a phyletic growth.

"The changes, which the individual organism experiences during its life in its material, physiological and morphological organization, are in part transmitted to its progeny. The changes thus acquired become more marked from generation to generation, until finally they result in a perceptible new structure."

"In this process, new or changing external influences undoubtedly exercise great activity, but the same influences, constantly repeated, must in the course of time also produce a change in the organisms through the physiological activity, which is conditioned by them, so that after a long time elapses, a species will have changed even in an unvarying environment and will react on new influences in a manner quite different from their progenitors; their "constitution" has undergone a change."

"This organic growth of living beings takes place regardless of the active use of the organs and in many cases remains independent of this (Lamarckian) factor of transformation. But use may exercise considerable influence on the formation resulting from the primitive organic growth, by modifying the growth, by restricting it to those parts most frequently called into use, or even by depriving other parts of the necessary matter (compensation)."

"The Lamarckian principle, therefore, offers but a possible and to transformation, the principal cause is to be found in organic growth."

"* * * The organic growth of butterflies is primarily conditioned by climatic influences. * * * The proof is to be found in the facts revealed by the geographical distribution of butterflies, by the variations corresponding to the seasons, and by experiments regarding the influence of artificial heat and cold on development."

Experimental proof is naturally of vital importance for Eimer's theory. He cites in this regard especially the experiments of Merrifield, Handfuss, Fischer, Fickert, and Countess Maria von Linden. In Eimer's own laboratory the latter performed experiments on Papilionides, "which prove in the most striking manner the recapitulation of the family-history in the individual." "The fact that it is possible by raising or lowering the temperature during the time of development to breed butterflies, possessed of the characteristics of related varieties and species living in southern and northern regions respectively, characteristics not merely of color and design, but also of structure, is complete irrefragable proof of my views."

Eimer therefore belongs to the class of naturalists, like Wigand, Askenasy, Naegeli, and many others, who reject the purely mechanical trend of Darwinism and recognize an "immanent principle of development." He seeks the essential cause of evolution in the constitution of the plasm of organisms. This very analogy between the development of the family and that of the individual should, in fact, convince any one of this. If Eimer chooses to refer the analogy to "growth" and to designate the evolution of the whole animated kingdom as also a process of growth, there is, strictly speaking, no room for objection. However, there is here a danger, which he does not seem to have guarded against. To designate the whole process as a growth, as Eimer does, really explains nothing, but merely defines more clearly the status of the problem. For, what do we know of the so-called process of growth? In truth, nothing, so that very little is gained by referring evolution to organic growth; the problem remains unsolved.

The most important and correct part of Eimer's conclusion seems to be the establishment of definite lines of development. He has, in fact, permanently disposed of the Darwinian assumption of universal chaos in evolution, upon which good mother Nature could at will exercise her choice. Fortuitously initiated development is a condition sine qua non of Darwinism and Weismannism. For any one, who has studied the work of Eimer and still adheres to this fundamental error of Darwinism, there is no possible escape from the labyrinth into which he has allowed the hand of Darwinism to lead him.

If, on the one hand, Eimer recognizes the immanent principles of development, he, nevertheless, on the other hand, also accords due consideration and ascribes great efficacy to external influences; in fact, he represents them as perhaps the more essential factor. Climate, nourishment, etc., affect the inner structure, the plasm, transform it and thus produce variation which is transmitted to the progeny. But, however great may be the influence of environment, Eimer seems to overestimate it. Indeed, the analogy of "growth" should have led Eimer to a conception of the true relation between "internal" and "external" causes. Warmth, air, light, moisture and nourishment, are undoubtedly necessary factors in the process of growth, but they are only the conditions which render it possible, and not the causes which produce it. The latter are to be found in the individual organism itself. The conditions may be ever so favorable and well-adapted for growth, still the organism will not develop unless it bear within itself the power to do so. On the other hand, although it is hampered and may become abnormal, it will readily grow even in an unfavorable environment, as long as it retains its inherent vital force. The same is very likely true of the genealogical growth. Evolution took place in virtue of the power inherent in the developing organisms. But only when the environment was favorable and normal, did the evolution proceed favorably and normally, that is, toward the perfection of the animate kingdom.

It appears as if the internal principle of development were losing influence and significance with Eimer; but the ulterior reason for this is not far to seek. Whoever recognizes the validity of the internal principle of development, eliminates chance, that stop-gap of materialism, from evolution, and is lead at once to a supreme Intelligence which directs evolution. As soon as it comes in sight, however, certain persons take fright and turn aside or even turn back in order to avoid it. This was the case with Eimer, although perhaps in a lesser degree. This is sincerely to be deplored, since his theory would have gained in depth if he had but done full justice to the internal principle of development. For the same reason he seems to have attacked Naegeli's principle of perfection, another fact which is very much to be regretted. True, it is as anti-mechanical as it can be and hence has gained but few adherents; but it is based on truth nevertheless, and will some day prevail in the doctrine of Descent.

It is perfectly intelligible that the thought of "perfection" should not have occurred to Eimer or should have slipped his memory during his observations on butterflies. The fact however, reveals a one-sidedness which he could have avoided. When the notion of utility is rejected--and Eimer rejects it very emphatically in his discussions on mimicry--it is undoubtedly difficult to arrive at the concept of a perfecting tendency. This, however, can in no way mean that this concept should be entirely banished from nature, even as the notion of utility cannot be banished. Even if the coloration and design of the wings of the butterfly do not reveal utility, other characteristics certainly do reveal it. It is one of the fatal mistakes of Darwinism, that it fails to recognize the possibility of dividing the characters and qualities of organisms into two large groups, as I attempted to do with more detail, for instance, in my "Catechism of Botany." There I called them (p. 89) "Autochthon-morphological" and "adaptive-morphological characters." The former reveal no relation to utility, they are innate and distinguish the organism from other organisms; the latter can be explained by means of certain vital functions, hence they possess a certain utility and adapt themselves more or less to environment. The former are permanent, the latter changeable. Darwinians regard all the characters of organisms as useful, physiological, and adaptive. If they have been hitherto unable to make good this assumption, they appeal to our lack of knowledge and console themselves with the thought that the future may yet reveal the missing relations. The presence on plants and animals of any autochthon-morphological characters means death to Darwinism, because these can never be explained by means of selection and struggle for existence.

Eimer is too much inclined towards the other extreme; he does not admit the existence of adaptive-morphological characteristics. Viewed in this aspect, his repudiation of mimicry may perhaps also seem somewhat harsh and one-sided. In this narrowness of view must also be sought the reason for his complete repudiation of Naegeli's principle of perfection.

It is an incontrovertible fact that in the organic world there exists an ascending scale from the imperfect to the perfect. Every organism is indeed perfect in its own sphere and from its own point of view. But perfection with reference to things of earth is a very relative concept; many an organism which is perfect in itself, appears very imperfect when compared with others. If, then, there is a gradation of animals and plants from the lower to the higher, it is the task of the theory of Descent to explain this gradual perfection. The crude and aimless activity of Darwinian selection, which necessarily operates through "chance," can never explain this perfection, which remains, as far as selection is concerned, one of the greatest enigmas of nature. Far from solving the enigma, selection but makes it obscurer.

If, then, one refuses to recognize a directing creative Intelligence, whose direction produces this perfection, nothing remains but Naegeli's principle of perfection. The outer world with its influences can certainly not produce perfection, hence this power must lie within the organism itself. But when one has once brought himself to accept an immanent principle of development, it surely cannot be difficult to take the next step and ascribe to it the tendency towards perfection.

That Eimer does not take this step, is, to my mind, a mistake, which must be attributed to his one-sidedness, which, in turn, results from the fact that he generalizes too arbitrarily his observations on butterflies and the conclusions which he draws from them. Animals and plants certainly possess many characteristics which cannot be explained by means of his theory alone. The conclusion will probably be finally arrived at, that nature is inexhaustible and many-sided, even in the lines on which it proceeds to attain this or that end.

One thing, however, of primary importance is evident from the investigations of Eimer, namely the proof that the same lines of development may be entered upon from entirely different starting-points, and that the number of these lines is limited. This fact is of importance because it enjoins more caution in arguing from uniformity of development to family-relation, than has been usually employed since the days of Darwin. The method commonly employed is undoubtedly very convenient, but is somewhat liable to be misleading. Hence, if one wishes to establish the genealogical relationship of forms, nothing remains but to set out on the laborious path of studying the development of both; and even then it remains questionable whether the truth will be arrived at. However, he who concludes to relationship from a comparison of developed forms, is much less likely to arrive at the truth.

In one point Eimer concedes too much to Darwinism, in the matter of the famous fundamental principle of biogenesis, according to which an organism is said to repeat in its individual development the whole series of its progenitors. Although he does not enter upon a discussion of the principle, it is evident from one passage that he accepts it. One is inclined to think that his careful observations and experiments should have convinced him of the contrary. It appears to me, at least, that the abundant materials of his observations bear evidence radically opposed to the principle. During late years, the antagonism to it has been on the increase, and the day is not very distant when it shall have passed into history. It would certainly be a laudable undertaking to enter upon a thorough investigation of the actual basis of the principle.

CHAPTER VI.

In every disease, especially in a lingering one, there are times when life's flickering embers glow with an unnatural brightness. Hence, it would not be a all surprising if a similar phenomenon were to be observed in the case of dying Darwinism; for it cannot be doubted that its disease is chronic. It has, in fact, been dying this long time. Certain indications render it very probable that we are at present witnessing such a phenomenon, for to-day we behold once more a few naturalists stepping before the public in defense of Darwinism. We are desirous of presenting the present status of the Darwinian theory as objectively as possible, hence, since we have hitherto heard exclusively anti-Darwinian testimonies--as the nature of the case demanded--we shall now lend our attention to a Darwinian. The reader may then decide for himself whether this treatise should not still bear the title, "At the Death-bed of Darwinism."

The naturalist in question is the zoologist, Professor F. von Wagner. In the "Umschau" (No. 2, 1900) he published an article, "Regarding the Present Status of Darwinism," which is highly instructive and important in more respects than one.

We wish, in the first place, to call special attention to the following statements embodied in the article: "It is not to be denied that in serious professional circles the former enthusiasm has considerably decreased and a scepticism is gaining ground more and more, which betrays a widespread tendency towards revolutionizing current theories. The fin de siecle therefore, finds Darwinism not with the proud mien of a conqueror, but on the defensive against new antagonists." And again: "It seems, in fact, as if Darwinism were about to enter a crisis, the outcome of which can scarcely be any longer a matter of doubt."

To what outcome reference is made, appears from two sentences in the Introduction: "Thus it happens that a theory which was once accorded enthusiastic approval, is treated with cold disdain or vice versa. Examples of this are to be found in the history of all sciences and circumstances seem to indicate that Darwinism is to add another to the number of these theories."

Is not this exactly what we have repeatedly asserted? It is most significant that these words are not written by an opponent of Darwinism, but by one who seems to be thoroughly convinced of the truth of Darwinism. I am of opinion that it can be no longer a matter of doubt to any one, that the position of Darwinism is hopeless. If this were not true, a Darwinian would be very careful about making such an open and unreserved statement.

We therefore accept Professor von Wagner's words as a very welcome endorsement of what we have constantly maintained. Professor von Wagner, however, proposes to himself the further question: Whence comes the unfavorable attitude of present-day natural science towards Darwinism? A discussion of this question by a Darwinian cannot but be of interest to us, and indeed is an important contribution to the problem. With Goette, Professor von Wagner admits that the objections, which are raised against Darwinism to-day, are the very same which were raised from thirty to forty years ago. But when he then proceeds to assert that this is not to be explained on the assumption that the pristine enthusiasm for selection was due to a serious over-estimation of that theory, he fails to furnish even a shred of evidence in support of his assertion.

Anyone can readily point out that Darwinism explains the totality of the world of organisms by interlinking them, but has generally failed to account for the individual case, Wagner admits this as far as the "actual" is concerned, for it is quite impossible to trace with any certainty the action, in any particular case, of natural selection in the process which results in the production of a new species. At the outset it was reasonable to hope, that with the progress of science this difficulty would be solved or at least lessened; but this expectation has not been realized. * * * It is wholly unintelligible how a naturalist can make this statement five hundred years after Bacon of Verulam, without drawing therefrom the proper conclusion. This lack of logic reminds me strongly of the assertion recently made by an eminent authority, that the principal cause of the difficulties of many naturalists in matters of religion is their deficient philosophical training.

Wagner's statement implies that, in the case of Darwinism one may in defiance of all established law, actually reverse the methods of natural science. How justifiable and how necessary was it not, then, that even three decades ago Wigand should have written his comprehensive work: "Darwinism and the Scientific Researches of Newton and Cuvier."

Ordinarily the scientific (inductive) method proceeds from the "actual" and attempts to deduce from the "individual case" an explanation, which applies to the whole. Here, however, we are face to face with a theory, which, according to the candid confession of an advocate, fails in the individual case, but furnishes a unifying explanation of the whole. This means nothing less than a complete subversion of all scientific methods. Usually a theory is deduced from separate observations regarding the "actual" but here--and this is what Wigand constantly asserted--the theory was enunciated first, and then followed the attempt to establish it in fact. One could then rest content and trust to the future to establish the theory by producing evidences of the "actual" in the individual case. But forty years have elapsed since the Darwinian hypothesis first became known, naturalists by the thousands have spent themselves in the endeavor to corroborate it by proofs based on actual facts, and to-day one of its own advocates has to confess that the endeavor has been a total failure. Instead of drawing the conclusion, however, that the theory is unwarranted and that the decrease of enthusiasm for it is therefore a natural consequence, he gratuitously enters a flat denial of this inference.

Every intelligent observer must conclude with absolute certainty from this confession of a Darwinian, that Darwinism is, in fact, not a scientific but a philosophic theory of nature.

But let us proceed to a consideration of the other reasons which Wagner suggests as an explanation of the retrogression of Darwinism. He states as a first reason, that scientific research since Darwin "has amassed such an abundance of empiric materials for the truth of the principle of Descent, that this doctrine has been able, even for some time past, to maintain an independent position and to draw proofs of its truth immediately from nature itself, without the intervention of Darwinism." * * * "From which it follows as a matter of course, that the question, whether the manner indicated by Darwin for the origin of species is the correct one, has decreased by no means inconsiderably in significance, inasmuch as Darwin's theory could now, if it were necessary, be abandoned with less concern than formerly because it could be relinquished without detriment to the doctrine of Descent."

It is unintelligible how one can attempt to explain a fact of such importance so superficially. With naive unconcern there appears on the face of it the acknowledgement that Darwinism has really not been based on actual observation but has been enunciated for the sake of the doctrine of Descent. Come what may, this must be vindicated. Other means are now said to substantiate it, hence the Darwinian crutches may safely be discarded. The principle of action twenty or thirty years ago was therefore: a poor explanation is better than no explanation. I cannot understand, how Wagner dares to credit present-day naturalists with such motives.

When he then proceeds to say "that with the advance of the principle of development, new lines were entered upon, which led primarily to the corroboration and empiric demonstration of the doctrine of Descent, and not of Darwinism"--that the theory of Darwin was consequently neglected and, in fact, forced into the background--"that the labors specifically attributable to Darwinism as compared with the theory of Descent, put the former more and more into a false position to the detriment of its prestige"--when, I say, Wagner has marshalled all these considerations to explain the present aversion to Darwinism, he is guilty of a total subversion of facts. The true state of the case is the very contrary.

The credit given by Wagner to the Darwinian theory for stimulating research, is the very same as I also accorded it. The purpose of this research undoubtedly was to substantiate not only the doctrine of evolution in general, but also the Darwinian hypothesis in particular. To verify this, one need only glance over the various numbers of the "Kosmos," the periodical, which Haeckel and his associates established for that very purpose and which continued to publish good and bad indiscriminately until some time in the eighties when lack of interest compelled its discontinuance. Wagner therefore misconstrues facts when he asserts that there have been no specifically Darwinian researches. Since the thoughts of Darwin first found expression these researches have been most abundant and their results have been consigned to the printer's ink. No doubt--and this is the salient point, which Wagner passes over in complete silence--they have been of service only to the doctrine of Descent in general, and in spite of the energetic efforts of the Darwinians, they have never led to the ardently desired proof from facts of the hypothesis of selection. This and no other is the state of the case.

In view of these vain endeavors, however, intelligent investigators have gradually become perplexed and have turned away from Darwinism, not because they have lost interest in it nor even because they no longer feel the need of it to assist the doctrine of Descent, but for the one sole reason that its insufficiency has become more and more apparent and that all experiments undertaken on its behalf have made the fact clearer and clearer that the first criticism of the great naturalists of the sixties and seventies was perfectly justified.

In forming a judgment concerning the whole question it cannot but be a matter of the utmost significance, that men have turned away from Darwinism to entirely different theories of Descent. It is a mistake to suppose, as Wagner would have us suppose, that the last decades have produced nothing but generalities regarding the doctrine of Descent. For they have also witnessed the publication of a number of significant works, which aimed at giving a better individual explanation than was found in Darwinism. I need but recall Naegeli, Eimer, Haacke and a host of others. The most noteworthy feature of these new views regarding theories of Descent, is the constantly spreading conviction that the real determining causes of evolution are to be sought for in the constitution of the organisms themselves, hence in internal principles. This view, however, is not only absolutely and diametrically opposed to Darwinism but completely destructive of it as well.

The actual circumstances, therefore, are the very reverse of those pictured by Wagner. Darwinism has been rejected not on account of a lack of research but on account of an abundance of research, which provided its absolute insufficiency.

Besides these "general points of view," as he calls them, Wagner finds two other "considerations of no less importance" for explaining the decay of Darwinism. It is an incontrovertible fact, that the hereditary transmission of acquired characters has in no way been proved. On the contrary after it had at first received a general tacit recognition and was postulated by Lamarck, Darwin and Haeckel, it was denied by Weismann. Wagner asserts "that the number of those who have allied themselves with Weismann in this matter is obviously on the increase as is naturally the case, since, to the present day not a single incontestable case of hereditary transmission of acquired characters has been demonstrated, where as actual facts are at hand to prove the contrary."

It is perfectly evident that the doctrine that acquired characters are not inherited is fatal to Darwinism. Hence Wagner rightly considers its ascendancy a notable factor in bringing about the decay of Darwinism.

Finally, Wagner briefly indicates that certain new theories necessarily exercised an influence on Darwinism. Haeckel and the palaeontologists of North America supplemented it with a number of Lamarckian elements without alteration of its essential principles (the Neo-Lamarckians); Eimer regards the transmission of acquired characters as an established fact, but rejects natural selection as wholly worthless; Weismann, on the contrary, denies the transmission of acquired characters, but nevertheless regards natural selection as the main factor in the formation of species (the theory of the Neo-Darwinians). Eimer speaks of the impotence of natural selection, Weismann of its omnipotence. All this has shaken men's confidence in the trustworthiness of the Darwinian principles. This fact we are in no way inclined to doubt, but we must again differ from Wagner with regard to its significance. We maintain that matters had to take this turn, since the reason why Darwinism is now meeting with such serious opposition, is to be found in its very nature. This indeed should have been recognized forty years ago instead of just beginning to dawn on men of science at the present day. For if acquired characters are not transmitted by heredity, Darwinism is an impossibility. Forty years ago Darwinism should have recognized that its first and supreme task was to prove the hereditary transmission of acquired characters, so as to establish itself, first of all, on a sound footing.

One of the most peculiar incidents in this scientific tragi-comedy is the fact that Weismann, the mainstay of contemporary decadent Darwinism, attacks with might and main its fundamental assumption, the transmission of acquired characters, whereas Eimer, who is thoroughly convinced that he has proved that doctrine, in his turn attacks Darwinism and proves with telling effect the impotence of its principles. The amused observer can really demand nothing more. He can but rub his hands for joy and cheer on the heated combatants: Well done! On with the struggle! and the last vestige of Darwinism will soon have disappeared.

If, then, we were to summarize our strictures on the reasons which Wagner adduces to account for the decay of Darwinism, we would say this: Some of them are unwarranted, others are falsely interpreted.

There is, however, a third point which is of special interest to us, in the article under consideration; we refer to the view, which there finds expression, regarding the nature and outcome of the present crisis--a crisis, which, as a candid naturalist, Wagner is not in a position to deny.

This view rests on the entirely gratuitous assertion, "that the decline, in the esteem enjoyed by Darwinism, is not due to a better insight arising from widened experience, but is primarily the expression of a tendency--a tendency which resulted almost as a psychological necessity from the precarious position into which Darwinism was forced under the sway of the theory of Descent." This assertion rests, as stated above, on wholly erroneous assumptions. It is a serious mistake, to speak in this connection of tendencies and even to brand them as a "psychological necessity." The decline in esteem is essentially due to experience, and indeed to experience which has made it certain that Darwinism has everywhere failed.

The importance of the present crisis in Darwinism is to be restricted even further, according to Wagner, by the fact, "that the real objections, urged against the theory of Darwin, are almost in every instance based on theoretic considerations, the validity of which can be put to the test only in fictitious cases. This manner of proceeding manifestly leads to the inevitable consequence, that the results thus obtained can claim no decisive weight against Darwinism. A decisive critique can be constructed only on the basis of experience, and in this connection it cannot be emphasized sufficiently, that, as yet, the path to it has been scarcely indicated, to say nothing of its having been actually pursued." The reason for this fact according to Wagner, is to be found "in the numerous and most extraordinary difficulties that arise in the way of the empiric investigation of the theory of selection."

After we have read all this, we instinctively ask ourselves: do we actually live at the beginning of the 20th century? Is it possible, that even at this late day the whole structure of scientific method is to be subverted in this fashion?

Just consider for a moment, what according to these words is the actual import of the whole article: Darwinism is a unifying explanation of the origin of the totality of the world of organisms, but fails in the individual case; in any specified case it is "almost impossible" to trace with any certainty the action of natural selection in the process which results in the production of a new species; that is, Darwinism was enunciated with a complete disregard for inductive method, as an hypothesis to explain the whole, and without actual proof in the concrete--a most unscientific procedure. Immediately after, however, the adversaries of Darwinism are asked in all seriousness to produce individual facts in disproof of the theory.

In the same strain Wagner goes on to say that "from no point of view is our vision so penetrating as to be able to grasp the coherence which according to Darwin pervades the complex course of natural selection. When men of science take occasion to repudiate Darwinism because of our inability to explain satisfactorily any particular case by means of the theory of selection, this inability arises not from the theory of Darwin but from the inadequacy of our experience. For as yet the empiric prerequisites for an objective judgment regarding the validity or futility of the theory of selection are entirely lacking." Every naturalist who believes in the inductive method must needs draw the conclusion from these naive admissions, that, as Darwinism lacks the empiric prerequisites, it should be discarded. Moreover, the demand is made in all seriousness, that, in order to refute Darwinism which has not as yet been established empirically, empiric proofs should be forthcoming.

To my mind, the scientific and logical bankruptcy of Darwinism was never announced more bluntly and ingenuously. Furthermore it must be remarked that Wagner's statement, regarding "fictitious cases," is not even pertinent. He seems to have no idea of the observations and experiments of Sachs, Haberlandt, Eimer, and a host of other investigators. The disproof of Darwinism on the basis of scientific research is an accomplished fact.

A word about the conclusion of Wagner's article, which in view of what has been already said, cannot be a matter of surprise. He maintains that the considerations which he adduces, "clearly" prove that there is no "reasonable ground for despairing of the theory of Darwin--; for a theory, which neither proceeds from questionable assumptions, nor loses itself in airy hypotheses, but rests throughout and exclusively on facts, need never fear the advance of science."

But a moment ago it was asserted that the theory of selection is lacking "entirely as yet the empiric prerequisites" and now only twenty-three lines further on, it rests "throughout and exclusively on facts." It is difficult to know what conclusion to come to regarding a naturalist and University professor who can commit himself to such a contradiction. I shall abstain from any comment and let the reader form his own judgment.

Does this article betoken the death-bed of Darwinism? For my own part I repeat what I said above, that I consider it the most valuable contribution to the characterization of decadent Darwinism that has appeared up to the present time. The sooner a theory, which is thus treated and characterized by one of its own advocates, is stored away in the lumber-room of science, the better. In view of the sound judgment, which is to-day becoming more and more apparent in scientific circles, there is reason to hope that this article of Professor von Wagner will be additional incentive for many naturalists to break completely with Darwinism.

CHAPTER VII.

In the year 1899 Haeckel published a new work, which he intended as a kind of testament; for with the close of the nineteenth century the author desired to put a finishing touch to his life-work.

In the Preface Haeckel states with very remarkable modesty that his book cannot reasonably claim to present a complete solution of the riddles of existence; that his answer to the great questions can naturally be only subjective and only partly correct; that his attainments in the different branches is very unequal and imperfect; and that his book is really only a sketch book of studies of very unequal value. In this way the author naturally gains at once the confidence of his reader who is thus prepared to yield assent when the author makes pretense to sincerity of conviction and an honest search after truth. The reader's surprise at the contents of the book and at the manner of its presentation is, however, only increased by this ruse. All modesty has vanished, monistic doctrines are presented as absolute truth, every divergent opinion is contemptuously branded as heretical; in short, the book reveals a Darwinian orthodoxy of the purest type, with all the signs of blind bigotry and odious intolerance which the author imagines he discovers in his Christian adversaries. It is difficult to see where, in view of such a contradiction between the work and its Preface, there is room for an honest striving after truth. Personally I do not wish to deny Haeckel all honesty of purpose, for it is my endeavor to understand the whole man. The one prominent feature of the "Weltraetsel" is the fact that, owing to a very marked deficiency in philosophical training, Haeckel has become so completely absorbed in his system that he has lost all interest in everything else and takes cognizance only of what suits his purpose. What he lacks above all, is the ability to appreciate even the "honest" opinion of others; hence, from the very outset he brings into the discussion that bitterness of which he complains in others (in the Weltraetsel he once makes this accusation against me). Notwithstanding all this, honest conviction may be present, but if so, it is joined with total blindness. But what is to be thought of his search after truth since he completely ignores his adversaries? For instance, in spite of Loofs' attacks, he continues to have his book reprinted without alteration, without submitting it to revision. The "Reichsbote" is perfectly in the right when it says: Haeckel, in fact, takes account only of what suits his purpose.

As regards the contents of the "Weltraetsel," it is not my intention to enter here upon a criticism of it but merely to discuss it as illustrating the general status of the theory of Descent. It is to be noted, in the first place, that it is really not a scientific book at all; for of its 472 pages, the first or "Anthropological Part," with which alone we are here concerned, occupies only 74 (from pages 27 to 100), even less than one-sixth of the whole, whereas the "Theological Part" is almost twice as long. The book is, in fact, rather a theologico-natural-philosophical treatise than a work of natural science. The scientific part is, however, the foundation on which Haeckel builds up his natural philosophy, and which he uses as the starting point of his criticism of theology. Hence it is worth our while to discuss it.

How then fares it with the anthropological basis of Haeckel's whole system? As an attentive student of his age the naturalist-philosopher of Jena must have perceived the true position of Darwinism, namely, that the foremost naturalists of to-day have no more than an historical interest in it. Since, in accordance with the well known tendency of old men to persevere in the position they have once assumed and not easily to accept innovations, Haeckel is still an incorrigibly orthodox Darwinian, we should naturally expect him to embody in this testament some new cogent evidence of the truth of Darwinism. But nothing of that nature is to be found in the book.

The first chapter of the "Anthropological part" is taken up with a "general history of nineteenth century culture," in itself a sign of peculiar logical acumen, that he should include this and the "struggle regarding world-views" in the "anthropological part" instead of embodying it in a general introduction. The remaining chapters treat: "Our Bodily Structure," "Our Life," "Our Embryonic-history," "Our Family-history." It is not to be supposed, however, that any arguments are here adduced, nothing but assertions; a large part of the chapter is taken up with historical sketches, in which Haeckel again proves himself utterly devoid of all appreciation of history and all sense of justice. He attributes the decay of the natural sciences to the "flourishing condition of Christianity" and dares to speak of the unfavorable influence of Christianity on civilization. Apart from the historical sketch, each chapter presents only the quintessence of Darwinism, fairly bristling with assertions, which are boldly put forth as incontrovertible truths. In view of the author's demand to have at least his sincere love of truth recognized, we can but throw up our hands out of sheer astonishment. To illustrate Haeckel's "love of truth" let it suffice to observe that in the second chapter he asserts that man is not only a true vertebrate, a true mammal, etc.--which indeed is passable--but even a true ape (having "all the anatomical characteristics of true apes"). With a wonderful elasticity he passes over the differences. What, indeed, is to be said, when he states as a "fact" that "physiologically compared (!), the sound-speech of apes is the preparatory stage to articulate human speech." It is so simply monstrous, that even Garner's famous book of ape-speech, cannot surpass it. As a third illustration of Haeckel's method of argumentation, if we are still justified in speaking of such a thing, we may mention his assertion (p. 97) as a "certain historical fact," "That man is descended directly from the ape, and indirectly from a long line of lower vertebrates." If, in view of the results of research during the last forty years any one can assert this as a "certain historical fact" and can still wish to be credited with honest conviction and love of truth, there remains, to adopt Haeckel's own expression, but one explanation for this psychological enigma, namely, intellectual marasmus senilis, which may very easily have set in with a man of sixty-six, who himself complains (p. 7) of "divers warnings of approaching age."

Thus, the anthropological part of the "Weltraetsel" contains nothing new; always the same old story, the same threadbare assertions without a shred of evidence to corroborate them.

The remaining parts also contain various scientific assertions, which are proposed as facts without being such, but these parts do not immediately pertain to our theme. Suffice it to say that, after reading Haeckel's "Weltraetsel," one would be led to think that there is no question of a "deathbed of Darwinism," but that on the contrary Darwinism, as remodeled by Haeckel, is more in the ascendant to-day than ever. Let us judge of its prestige by the reception accorded the "Weltraetsel."

One unaltered edition after the other, thousand after thousand, the book is given to the public. Hence it must meet with approval. It does indeed meet with approval, but the question is, from whom? Immature college and university students will doubtless receive it with reverential awe, just as they received the "Natural History of Creation" twenty-five years ago. Bebel accepts the book as an infallible source of truth, and after him the social democrats and free-church members will add it to the list of their "body and stomach books," which alone will afford it a respectable clientele, at least in number. In no one of my "deathbed articles," however, have I as yet ever maintained that Darwinism was decadent in these circles. I know full well, that Darwinism has filtered down into that sphere and there satisfies the anti-Christian and anti-religious demands of thousands.

Nothing, however, really depends on these senseless blind adherents of Haeckel's unproved assertions. We are now intent upon investigating how the world of eminent thinkers and natural science regards the latest product of Haeckel's fancy. That alone is of importance in ascertaining the real status of Darwinism.

As regards, in the first place, the other parts of the book, it is well known that all of them were vigorously attacked. Loofs in particular exposed Haeckel's theology, according to its deserts, in the clear light of truth, and convicted Haeckel of "ignorance" and "dishonesty;" while the philosopher Paulsen made short work of the "Weltraetsel" from his own standpoint, ("if a book could drip with superficiality, I should predicate that of the 19th chapter"). Harnack also condemned the theological section in the "Christliche Welt," and Troeltsch, Hoenigswald, and Hohlfeld took Haeckel severely to task on philosophic grounds. The naturalists have thus far maintained silence.

Scientific journals, and, I believe, only the more popular ones, pass a varying judgment on the book according to the intellectual bent of their book reviewers; but no one of the eminent and leading naturalists has publicly expressed his opinion regarding it. They all maintain a very significant silence, which speaks for itself. Now, however, just at the proper time a book, Die Descendenztheorie has appeared from the pen of the zoologist, Professor Fleischmann of Erlangen, in which Haeckel is severely condemned. (See Chapter IX.)

The press-notices of the Weltraetsel, which are quoted in the book will be considered presently. It appears that with reference to natural science, only "laymen" discuss the book and approve of Haeckel's views. This is a point of great importance since it proves satisfactorily that men of science will have nothing to do with the "Weltraetsel." The large number of replies would, however, not allow Haeckel's friends to remain silent. The most extensive defense forthcoming was a pamphlet published by a certain Heinrich Schmidt of Jena. It cannot be gathered from his book (Der Kampf um die Weltraetsel, Bonn, E. Strauss 1900) to what profession the author belongs, hence I am unable to judge whence he derives the right to treat Haeckel's opponents in summary a manner. It is significant to note what class of men, according to Schmidt, received the "Weltraetsel" with enthusiasm and joy. They are August Specht, the free-church editor of "Menschentum" and of the "Freien Glocken," Julius Hart, Professor Keller-Zuerich, the philosopher and "Neokantian" Professor Spitzer of Graz, the popular literateur W. Boelsche, W. Ule, and a few unknown great men, Dr. Zimmer, Th. Pappstein, R. Steiner, A. Haese; but stay, I came very near forgetting the great pillar, Dodel of Zuerich. But where is there mention of the professional colleagues of Haeckel whose testimonies could be taken seriously? Under the heading "Literary Humbug," which evidently has reference to the contents of his own work, Schmidt then meets numerous objections. Here vigorous epithets are bandied about, as, for instance, "absolute nonsense," "muddler," "foolish and senseless prattle," "idle talk," etc.; and from Dodel he copies the words with which the latter once sought to annihilate me: Job, verse 10, "Thou hast spoken like one of the foolish women." And he ventures to express indignation at Loofs' "invectives." As a compliment to Lasson he declares that he could easily conceive of the possibility of an ape ascending the professor's chair and speaking as intelligently as he (Lasson); which remark he probably intended as a witticism. He informs his readers that the criticism of Haeckel by men like Virchow, His, Semper, Haacke, Baer, and Wigand have been examined by professional specialists and proved practically worthless. This statement alone so clearly reveals Schmidt's lack of critical faculty and judgment that by it he at once forfeits his right to be taken seriously.

The whole book is nothing more than a collection of quotations from the reviews of the "Weltraetsel," interspersed with characteristic expressions like "idle talk," "nonsense," etc., as exemplified above. A really pertinent reply and refutation of objections is entirely beyond Schmidt's range; he waives the demand for a direct reply, for instance, in the following amusing way (p. 28): "Two reasons, however, prevent me from being more explicit: In the first place I do not like to dispute with people who adduce variant readings and church-fathers as proofs and can still remain serious. In the second place I would not like to fall into the hands of a Loofs." In this manner it is indeed easy to evade an argument, which for good reasons one is not able to pursue. Loofs' criticism is so serious and destructive that it should be of the utmost concern to Haeckel's friends to refute it. Since they are unable to do so, they content themselves with references to Loofs' caustic style, which he should indeed have avoided. There are, nevertheless, cases in which one must employ trenchant phraseology, and Haeckel himself has given an occasion for it; a dignified style is simply out of the question in his case. Haeckel extricated himself with even greater ease, by declaring that he had "neither time nor inclination" for reply, and that a mutual understanding with Loofs was impossible because their scientific views were entirely different. Could anything be more suggestive of the words of Mephistopheles:

"But in each word must be a thought-- There is,--or we may so assume,-- Not always found, nor always sought. While words--mere words supply its room. Words answer well, when men enlist 'em, In building up a favorite system."

There are two other points in Schmidt's book that are of interest to us. The first of these is the manner in which the author treats the Romanes incident. Romanes ranks, as is well known, among the first of Haeckel's authorities. Hence it is a very painful fact that, but a short time before the publication of the first edition of the "Weltraetsel," my translation into German of Romanes' "Thoughts on Religion" should have appeared. From this book it was evident that Haeckel and his associates could no longer count this man among their number since he--a life-long seeker after truth--had abandoned atheism for theism, and died a believing Christian. Troeltsch and the "Reichsbote" asked whether Haeckel had purposely concealed this fact, and Schmidt now explains that Haeckel first became acquainted with the "Thoughts on Religion" through him towards the end of January, 1900. Unfortunately he does not add that since then a number of new editions of the "Weltraetsel" have appeared, in which Haeckel could have explained himself in an honorable manner. Schmidt has therefore not been successful in his attempt to clear up this matter.

But how does he settle with Romanes? He says: "We are assured that the thoughts were written down by the English naturalist George John Romanes"; and again: "The thoughts are published by a Canon of Westminster, Charles Gore, to whom they are said to have been handed over after the death of Romanes in the year 1894." Then he has the audacity to place Romanes in quotation marks. And finally he asserts that they would abide by Romanes' former works as their authority, the more so, because these were not, like the "Thoughts," "published and glossed by a Canon only after his (Romanes') death." By means of all this and of a comparison with the "Letters of the Obscurantists" he wishes to create the suspicion that there might be question here of forgery. Such an insinuation, (I employ Schmidt's own words) "cannot be characterized otherwise than as contemptible." "Here it is even worse than contemptible." I must beg my reader's pardon for overstepping the bounds of reserve with these caustic words, although they originated with Schmidt; but really the flush of anger rightfully mounts to one's cheeks when a man, from the mere fact that he is a disciple of the "great" Haeckel assumes the right to charge Canon Gore and indirectly myself with forgery. It is really very significant that these men should have to resort to such base and despicable expedients to extricate themselves from their unpleasant predicament. Apart from this, it was very amusing to me personally to think that for the sake of my unworthy self, Schmidt should have borrowed from his lord and master the epithet "pious," which Haeckel in his turn has drawn from his cherished friend Dodel. In all probability they will continue to hawk it about in order to bring me into disrepute with the rest of their kind. The few remarks Schmidt still finds it proper to make regarding the "Thoughts," betray his inability to understand the book. But as I stated in the preface it was a difficult book to read and understand. It is obviously not reading matter for shallow minds. I refer Schmidt to the biography of Romanes, published by his wife, (The Life and Letters of G. J. Romanes, London, Longmans, Green & Co., 1898), where he will find Romanes' religious development described by a well-informed hand. This development began as early as 1878, hence during the time of his intimate friendship with Darwin. In this book on pages 372 and 378 Schmidt will also find the words in which, before his death, Romanes begged that, if he were personally unable to publish the "Thoughts," they should be given to his friend Canon Gore after his own death. But why waste so many words on Mr. Schmidt, for since all these things must be doubly disagreeable and painful to him and Haeckel, he will very probably resort without delay to personal insinuation and accuse Mrs. Romanes of forgery.

To us, however, who thoroughly appreciate the situation, it is a matter of great moment that of one of the few really eminent naturalists, to whom Haeckel thought to be able to lay full and exclusive claim, for the last twenty years of his life should have been moving towards the Christian faith in his eager search for truth and should die not a monist, but a convinced Christian. Neither did he die an old man, to whom the adherents of monism would certainly have the effrontery to impute feeble-mindedness, but at the early age of forty-six years. Nor was his a sudden deathbed conversion--an impression which Schmidt attempts to create (p. 62) in order to be able with H. Heine to relegate the conversion to the domain of pathology--but followed after many years of diligent and honest study and research. The other point of which we must treat here, is the manner in which, after the example of Dr. Reh, Schmidt attempts in the "Umschau" to exonerate Haeckel in the matter of the "History of the three cliches." To begin with, it is at the very least dishonest on the part of Schmidt to say that, "in default of scientific arguments, theological adversaries have for the last thirty years been using it as the basis of their attacks." That is untrue, the "theological adversaries" have not had knowledge of it for that length of time. On the contrary Haeckel's own scientific colleagues were the first to discover and publish the matter some time in the seventies, and in consequence excluded Haeckel from their circle. Why does Schmidt not mention here the names of Ruetimeyer, His, and Semper? Furthermore Schmidt writes as if Haeckel had satisfied his colleagues in the matter of his forgery by declaring soon after (1870) that he had been "guilty of a very ill-considered act of folly." Why does Schmidt not mention the fact that the weighty attacks of His (Our Bodily Form and the Physiological Problem of its Origin, Leipzig, 1875) dates from the year 1875, five years after Haeckel's forced, palliative explanation? Besides, this incident of the three cliches is only one instance; the other examples of Haeckel's sense of truthfulness are for the most part entirely unknown to his "theological adversaries," who have nowhere to my knowledge made use of them; but all of them have been brought to light and held up before Haeckel by naturalists, namely, by Bastian (1874), Semper and Kossmann (1876 and 1877), Hensen and Brandt (1891), and Hamann (1893). Does this in any way tend to establish Schmidt's honesty? (Dr. Dennert has entered into a more searching criticism of Haeckel in his book, Die Wahrheit ueber Haeckel. 2 Aufl Halle a. S., 1902.)

In a word, the manner in which the "Weltraetsel" was received and in which Haeckel has been defended by Schmidt, are valuable indications of the decay of Darwinism. I repeat that I am speaking of course of the leading scientific circles. Those who hold back are never lacking, and one cannot be surprised that, in the case of Darwinism, their number is considerable: for on the one hand, to understand it an extraordinarily slight demand is made on one's mental capacity; and on the other hand it is a very convenient and even a seemingly scientific means of obviating the necessity of belief in God. These facts appeal very strongly to the multitude.

In concluding this section, we shall quote a positive testimony to the decay of Darwinism. On page 3 of his "Outlines of the History of the Development of Man and of the Mammals" (Leipzig, W. Engelmann, 1897) Prof. O. Schultze, Anatomist in Wuerzburg, says: "The idea entertained by Darwin, that the development of species may be explained by a natural choice--Selection--which operates through the struggle of individuals for existence, cannot permanently satisfy the spirit of inquiry. Even the factors of variability, heredity, and adaptation, which are essential to the transformation of species, do not offer an exact explanation."

CHAPTER VIII.

I have already called attention several times to the fact that Darwinism is indeed on the wane among men of science, but that it has gradually penetrated into lay circles where it is now posing as irrefragable truth. Especially the circles dominated by the social democrats swear by nothing higher than Darwin and Haeckel. In fact, only a short time ago Bebel publicly professed himself a convert to Haeckel's wisdom.

It is inevitable, however, that light should gradually dawn even in these circles, for it would be indeed strange, if no honest man could be found to tell them the truth regarding Darwinism. This has occurred sooner than I dared to hope. This chapter can announce the glad tidings that even in "social-democratic science" Darwinism is doomed to decay. Much printer's ink will, of course, be yet wasted before it will be so entirely dead as to be no longer available as a weapon against Christianity; but a beginning at least has been made.

In the December number of the ninth year of the Sozialistische Monatshefte, a social-democratic writer, Curt Grottewitz, undertakes to bring out an article on "Darwinian Myths." It is stated there that Darwin had a few eminent followers, but that the educated world took no notice of their work; that now, however, they seemed to be attracting more attention. "There is no doubt, that a number of Darwinian views, which are still prevalent to-day, have sunk to the level of untenable myths. True, the main doctrine of Darwin--the origin of new species from existing ones--is incontestably established, but apart from this even some very fundamental principles, which the master thought he discerned in the development of organisms, can scarcely be any longer maintained."

It may be well to remark here, that this was not really Darwin's main doctrine, for it already existed before his time (Lamarck, Geoffroy St. Hilaire). Darwin's main doctrine is the explanation of the origin of species by natural selection operating through the struggle for existence. It is therefore the old error repeated. Darwinism is confounded with the doctrine of Descent, of which it is merely one form. It is not our intention to derogate in the least from Darwin's merit, which consists in the fact that he gained general recognition for the doctrine of Descent; but that was not his main work. He wished above all to explain the How of Descent; this is his doctrine, and this doctrine we attack and declare to be on the point of expiring.

Grottewitz very frankly continues: "The difficulty with the Darwinian doctrines consists in the fact that they are incapable of being strictly and irrefutably demonstrated. The origin of one species from another, the conservation of useful forms, the existence of countless intermediary links, are all assumptions, which could never be supported by concrete cases found in actual experience." Some are said to be well established indirectly by proofs drawn from probabilities, while others are proved to be absolutely untenable. Among the latter Grottewitz includes "sexual selection," which is indeed a monstrous figment of the imagination. There was moreover really no reason for adhering to it so long. It is eminently untrue, that the biological research of the last few years proved for the first time the untenableness of this doctrine, as Grottewitz seems to think. Clear thinkers recognized its untenableness long ago, and surely Grottewitz and the whole band of Darwinian devotees as well, could have known that as early as twenty-five years ago this doctrine had been subjected to a reductio ad absurdum with classic clearness in Wigand's great work.

It is certainly a very peculiar phenomenon; for decades we behold a doctrine reverently re-echoed; thoughtful investigators expose its folly, but still the worship continues, the Zeitgeist must have its idol. It appears, however, as if the Zeitgeist were gradually tiring of its golden calf and were on the point of casting it into the rubbish-heap. Misgivings arise on all sides; here one class of objections are considered, there another. A closer examination reveals that these are by no means new reasons, based on new researches, but the very oldest, urged long ago and perhaps much more clearly and forcibly. At that time, however, the Zeitgeist was under the spell of the suggestion of individual men: it heard and saw nothing but the captivating, obvious simplicity of the doctrine; but now when the subject begins to be tedious and the discussion lags, the interest consequently abates and the Zeitgeist suddenly grasps the old objections, presented in a new garb, and what was hitherto truth, clear and irrefutable, now sinks into the dreary, gray mists of myth. Sic transit gloria mundi!

This has been the history of Darwinism, and especially of Darwin's theory of sexual selection. What Grottewitz urges against it, was advanced decades ago by other and more eminent men; then people would not listen, to-day they are inclined to listen. Of very special interest is the further admission, that "the principle of gradual development" has been "considerably shaken" and is "certainly untenable." Grottewitz points out that it has been demonstrated that the progeny of the same parents are often entirely dissimilar, and that new organs very suddenly spring up in individuals even when they had had no previous existence. "A slight variation from the parent form is of no utility to the progeny; they must acquire at once a completely developed, new character, if it is to be of any use to them." Quite right! but this one admission is destructive of the entire doctrine of natural selection. If one accepts saltatory evolution, as for instance, Heer, Koelliker, and Wigand did long ago, then, as Grottewitz now discovers, the difficulty arising for Darwinism from the absence of the numerous intermediary forms which it postulates, naturally disappears.

Grottewitz attributes sudden variation to the influence of environment, just as Geoffroy St. Hilaire had already done before Darwin. He likewise repudiates Darwin's doctrine of adaptation and the theory of "chance," which is bound up with all his views. "Darwin's theory of chance seems to me to be especially deserving of rejection." The article closed with these words: "There must evidently be a very definite principle, according to which the frequent and striking development from the homogeneous to the heterogeneous, from the no-longer adapted to the readapted, proceeds. We all of us are far from considering this principle a teleological, mystical or mythical one, but for that matter, Darwin's theory of chance is nothing more than a myth."

He is most certainly in the right. To place this whole wonderful, and so minutely regulated world of organisms at the mercy of chance is utterly monstrous, and for this very reason Darwinism, which is throughout a doctrine of chance, must be rejected; it is indeed a myth. We are grateful to Grottewitz for undertaking to tear the assumed mask of science from this myth and expose it before his associates. He should, however, have done so even more vigorously and unequivocally and should have stated plainly: Darwinism is a complete failure; we believe indeed in a natural development of the organic world, but we are unable to prove it.

In the conclusion of the article quoted there is, of course, again to be found the cloven-hoof: by all means no teleological principle! But why in the world should we not accept a teleological principle, since it is clearly evident that the whole world of life is permeated by teleology, that is, by design and finality? Why not? Forsooth, because then belief in God would again enter and create havoc in the ranks of the "brethren."

But however much men may struggle against the teleologico-theistic principle and secure themselves against it, it is all of no avail, the principle stands at the gate and clamors loudly for admission; and if Grottewitz could but bring himself to undertake a study of Wigand's masterful work, perhaps his heresy would increase and we might perhaps then find another article in the "Sozialistische Monatshefte" tending still more strongly toward the truth.

But what will Brother Bebel with his Haeckelism say to the present article?

All in all, instead of calling his article "Darwinian Myths" Grottewitz might just as well have entitled it "At the Deathbed of Darwinism." May he bring out a series of "deathbed articles" to disclose the truth regarding Darwinism to his associates.

CHAPTER IX.

Professor Fleischmann, zoologist in Erlangen, recently published a book bearing the title, "Die Descendenztheorie," in which he opposes every theory of Descent. The book is made up of lectures delivered by the author before general audiences of professional students, hence is popular in form and of very special apologetic value. Numerous excellent illustrations aid the reader in understanding the text.

One statement in the Introduction characterizes the decided position assumed by the author. He says: "After long and careful investigation I have come to the conclusion that the doctrine of Descent has not been substantiated. I go even farther and maintain that the discussion of the question does not belong to the field of the exact sciences of zoology and botany." At the outset, Fleischmann establishes the fact that in the animal kingdom there are rigidly separated types, which cannot be derived from each other, whereas the doctrine of Descent postulates "one single common model of body-structure" from which all types have been developed. Cuvier in his day, set up four such types of essentially different structure; when Darwin's work appeared two more had been added; R. Hertwig postulates even seven, Boas nine (both 1900); J. Kennel (1893) seventeen, and Fleischmann himself sixteen. In consequence the doctrine of Descent has become more complicated since it now embraces sixteen or seventeen different problems, each of which in turn gives rise to many subordinate problems.

The discussion which the author inaugurates regarding the domain to which the question of Descent belongs, is very well-timed. He forcibly and definitely discountenances the method which transfers it to the domain of religion. The question must be decided by the naturalists themselves according to the strict inductive method; that is, the solution must be based on well ascertained facts, without resorting to conclusions deduced from general principles. "Exact research must show that living organisms actually have overstepped the bounds defining their species, and not merely that they conceivably may have done so." Hence it is absolutely necessary to procure the intermediary forms. This is the foundation on which Fleischmann builds and against which no opponent can prevail. Fleischmann first discusses the differences between the classes of vertebrates; the mammals, birds, reptiles, amphibians and fish. For if the differences of their bodily structure could be shown to be one of degree and not radical, it could be supposed that the lines of demarcation which now delimitate the larger types might some day vanish. A single illustration suffices for Fleischmann's purpose, viz., the plan of structure of the limbs of the different classes of vertebrates. The four higher classes are characterized by a common underlying plan of limb structure, whilst fish have one peculiar to themselves. On the other hand it is an inevitable postulate of the doctrine of Descent that fish are the original progenitors of all other vertebrates. Hence the five-joint limbs of the latter must have developed from the fins of fish. This derivation was actually attempted but without success, as Fleischmann points out at considerable length. By means of citations taken from the writings of Darwinian adherents, he illustrates the confusion which even now reigns among them on this matter. The evolution of the remaining vertebrates from the fish is therefore a wholly gratuitous assumption devoid of any foundation in fact.

Fleischmann further discusses the "parade-horse" of the theory of Descent. It has been the common belief, especially fostered by Haeckel, that the history of the Descent of our present horse lies before us in its complete integrity as pictured in the drawings of Marsh. Here Fleischmann again proves at great length the insufficiency of actually available materials. Of special importance is his repeated demand that not only individual parts of the animals but the whole organism as well should be derived from the earlier forms. If, for instance, it be possible to arrange horses and their tertiary kindred in an unbroken line of descent according to the formation of their feet, whilst the other characteristics (teeth, skull-structure, etc.,) do not admit of arrangement in a corresponding series, the first line must be surrendered.

Very similar to this is the case of the "family history of birds," which as all know, has been traced back to reptiles. It is in this matter that the famous Archaeopteryx plays an important part. Unfortunately, however, grave difficulties are again encountered in this connection. This primitive form is a real bird according to Zittel; and according to the same investigator as also according to Marsh, Dames, Vetter, Parker, Tuerbringen, Parlow and Mehnert, it is inadmissible to connect birds with a definite class of reptiles. Haeckel finds his way out of the difficulty by supplying hypothetical forms which no one has ever seen, but which his imagination has admirably depicted as transitional forms. In so doing, however, he abandons the inductive method of natural science.

It is impossible for us to treat at such length all the remaining sections of this important book. We may mention in passing that Fleischmann examines the "roots of the mammal stock," and enters upon a detailed discussion of "the origin of lung-breathing vertebrates," the "real phylo-genetic problem of the mollusks," and "the origin of the echinodermata." It is evident that he boldly takes up the most important problems connected with the theory of Descent, and does not confine himself to a one-sided discussion of individual points. As he did not fear to examine thoroughly the famous, and as it hitherto appeared, invulnerable, "parade-horse," so neither does he hesitate to demolish the other reputed proof for the doctrine of Descent, e.g., the fresh-water snail of Steinheim, the remains of which Hilzendorf and Neumayr examined and were said to have arranged in lines of descent that "would actually stagger one." It is important to call especial attention to this because the adversaries of the book ignore it. He next shows up the so-called "fundamental principle of biogenesis" according to which organisms are supposed to repeat during their individual development the forms of their progenitors (enunciated by Fritz Mueller and Haeckel). Fleischmann points out the exceptions which Haeckel attributes to "Cenogenesis," (that is to falsification) and shows the disagreement among contemporary naturalists regarding this fundamental principle. Even Haeckel's friend and pupil, O. Hertwig sounds the retreat.

The 15th chapter deals with the "Collapse of Haeckel's Doctrine," which is revealed in the fact that "the practical possibility of ascertaining anything regarding the primitive history of the animal kingdom is completely exhausted and the hope of so doing forever frustrated." "Instead of scientists having been able from year to year to produce an increasing abundance of proof for the correctness of the doctrine of Descent, the lack of proofs and the impossibility of procuring evidence is to-day notorious." In the last chapter Fleischmann finally attempts to prove on logical principles the untenableness of the evolutionary idea.

He starts from the fact that philosophers use the word development to designate a definite sequence of ideas, i.e., in a logical order. "Metamorphosis, says Hegel, belongs to the Idea as such since its variation alone is development. Rational speculation must get rid of such nebulous concepts as the evolution of the more highly developed animal organisms from the less developed, etc."

Naturalists use the word in a different sense. Instead of a sequence of grades of being they posit a sequence of transformations; instead of a logical sequence of ideas they posit a transforming and progressive development. Zoology constructs a system of specific and generic concepts, "an animal kingdom with logical relations." Our concepts are derived from natural objects, but in reality do not perfectly correspond to them. The phylogenetic school commits the capital mistake of presenting a transformation which can be realized only in logical concepts, as an actually occurring process, and of confounding an abstract operation with concrete fact. "The logical transformation of the concept ape into the concept man is no genealogical process." The mathematician may logically 'develop' the concept of a circle from that of a polygon, but it by no means follows that the circle is phylogenetically derived from the polygon.

Because the concept of species is variable, the species themselves, according to Darwin, should be subject to a continual flux; whereas the real cause of the variability which he observed lies in the discrepancy between objective facts and their logical tabulation, in the narrowness of our concepts and in the lack of adequate means of expression. He thus makes natural objects responsible for our logical limitations.

With regard to organisms the Descent-school confounded the purely logical signification of the word "related" with that of blood or family affinity. But surely when they speak of the relation of forms in the crystal systems, they do not refer to genetic connection. To-day this interchange of concepts is so general that one needs to exercise great care if one would avoid it.

The theory which postulates the blood-relationship of individuals of the same species may be correct, but it is utterly incapable of proof, and the same is true in a greater degree when there is question of individuals of the same class but of different species. Since a direct proof is impossible, an attempt was made to construct an indirect proof by a comparison of bodily-organs. But in so doing the Descent theorizers had to relinquish scientific analysis altogether.

In conclusion Fleischmann states that he does not mean to discard every hypothesis of Descent. He simply gives warning against an over-estimation of the theory. In opposition to those who esteem it as the highest achievement of science, he looks upon it as a necessary evil. Its proper sphere is the laboratory of the man of science, and not the thronging market-place.

"The Descent hypothesis will meet the same fate (be cast aside), since its incompatibility with facts of ordinary observation is manifesting itself. At the time of its appearance in a new form, forty years ago, it exercised a beneficial influence on scientific progress and induced a great number of capable minds to devote themselves to the study of anatomical, palaeontological and evolutionary problems. Meanwhile, however, viewed in the light of a constantly increasing wealth of actual materials, the hypothesis has become antiquated and the labors of its industrious advocates makes it obvious to unbiased critics, that it is time to relegate it ad acta."

* * * * * * *

My own views agree with those of Fleischmann as presented above, except in regard to his last chapter. I must, of course, admit that his criticism has discredited the doctrine of Descent as a scientifically established theory. Hence, as I have always asserted, it must be excluded from the realm of exact science. No doubt people will come gradually to see that the theory involves a creed and therefore belongs to the domain of cosmic philosophy. All this I readily admit.

Not so, however, as regards the concept of "development." It seems to me to be incorrect to regard this as a logical concept only, even with reference to organisms. True, the whole zoological system is in reality nothing more than a logical abstraction. And in view of this fact one must be on one's guard against confusing a logical transformation of concepts with a genealogical development.

We must, however, not forget that we possess the wonderful analogy of ontogeny (individual development) and above all, the fact of mutation and of metagenesis. And even if we wish to avoid the error of Haeckel and others who find a necessary connection between ontogeny and phylogeny, nevertheless the analogy will still entitle us to picture to ourselves the development of the whole range of living organisms. Such a representation will, of course, have only a subjective value.

No doubt, it is logically unjustifiable to argue from the variable concept to the variability of the species. Still there is something real in plants and animals which corresponds to our specific concepts. In some cases the corresponding reality may be so well defined that it is not difficult to form the concept accurately; whereas in other cases where the task is more difficult, the difficulty must be due to the object. Under these circumstances we may safely conclude from the lack of definiteness in our concepts to a certain lack of rigid delimitation in the organic forms.

This blending of certain forms suggests the idea of transformation, but does not furnish definite proof of it. Such proof can be had only by the direct observation of a transformation. And no doubt in certain cases a transformation may occur. As regards animals, I may call attention, for instance, to the experiments made with butterflies by Standfuss, and as regards plants, to the experiments of Haberlandt, of which I treated in Chapter III. The limits within which these transformations take place are indeed very narrow as are also the limits of those indisputable varieties which naturally arise within an otherwise rigidly defined species. I am aware that the transformation of one species into another has not yet been effected, but the above-mentioned attempts at transformation have nevertheless demonstrated that certain organic forms when subjected to changed conditions of life, display certain mutations which clearly show that variability is to be attributed, not, certainly, to the specific concepts, but to the corresponding reality. This observation and reflexion, joined with the fact that organisms form a progressive series from the simple to the more complex, and with the observed phenomena of individual development, lead me to regard the concept of Descent as admissible, and in a certain sense, even probable. But I agree with Fleischmann in saying that this is a mere belief, and that all attempts to give it a higher scientific value by inductive proof have signally failed.

My standpoint, moreover, requires me to admit the validity of the hypothesis of Descent as an heuristic maxim of natural science. I believe that we shall be justified in the future, as we were forty years ago, in directing our investigation in the direction of Descent, and I do not consider such investigation so utterly hopeless as Fleischmann represents it. However, I entirely concur with him in the opinion that we are here concerned (and shall be for a long time to come) with a mere hypothesis which belongs not in the market-place, nor among the world views of the multitude, but in the study of the man of science.

Above all it must not be mixed up with religious questions. Whether the hypothesis will ever emerge from the study of the man of science as a well-attested law, is still an open question, incapable of immediate solution.

* * * * * * *

It is of interest for us to inquire what reception Fleischmann's protest against the theory of Descent has been accorded by his associates.

Fleischmann was formerly an advocate of the theory of Descent. He was a pupil and assistant of Selenka, who was then at Erlangen (died in Muenster 1902). He had previously written a number of scientific works from the standpoint of the Descent theory. In the year 1891, investigations regarding rodents led him to oppose that theory. During the winter term of 1891-92 he gave evidence of this change in a public lecture. Not until 1895 was there question of his appointment to the chair of zoology in Erlangen. In 1898 he published a Manual of Zoology based on principles radically opposed to the doctrine of Descent. This manual irritated Haeckel so much that he issued one of his well-known articles, Ascending and Descending Zoology, in which, after his usual manner, he casts suspicion on Fleischmann of having received his appointment to the chair at Erlangen by becoming an anti-Darwinian in accordance with a desire expressed at the diet of Bavaria. I am not aware that Haeckel has paid any attention to the work of Fleischmann which we have just reviewed.

By its publication, however, the author disturbed a hornet's nest. Dispassionate, but still entirely adverse is Professor Plate's review in the "Biologisches Zentralblatt," while the "Umschau" publishes two criticisms, one by Professor von Wagner, the other by Dr. Reh, which for want of sense could not well be equalled. It was the former who furnished material for our sixth chapter and who there displayed such utter confusion of thought regarding the inductive method. The same confusion is apparent in his recent utterance in which he observes that Fleischmann's whole aim is to accumulate observational data, meanwhile avoiding speculation as far as possible. His criticism is replete with bitter personal epithets, e.g., "reactionary," "mental incompetency," "dishonest mask of hypercritical exactness," which manifest the writer's inability to enter upon an objective discussion of the question.

A still more reprehensible position is assumed by Dr. Reh, who censures Fleischmann for introducing to the general public the question of Descent which belongs properly to the forum of science. He claims that Fleischmann, by so doing, forfeited his right to an unbiased hearing. Dr. Reh forgets that but a short time ago he had no word of censure for Haeckel's Weltraetsel which was intended for a far wider circle of readers. He next appropriates Haeckel's suspicion regarding Fleischmann which we noticed above, and then adds the entirely untrue assertion that the first half of Fleischmann's Manual, written before he took possession of the chair in Erlangen, is written in the spirit of Darwin, whereas the second half which appeared at a later date is written in the contrary spirit. He then takes individual points of Fleischmann's treatise out of their context in order to execute a cheap and nonsensical criticism of them. Haeckel has evidently been giving instructions on the best manner of dealing with adversaries. And very docile disciples they are who imitate his method even to the extent of defaming and abusing their scientific opponents.

But is not this another plain indication of the decay of Darwinism? Of course Haeckel recognized at the very beginning of his career that it was necessary to support the theory by means of personal bitterness, forgeries and misrepresentations. But if the last surviving advocates of Darwinism must needs have recourse to the same disreputable means, to what a low estate, indeed, has it fallen!

Let us hope that these last wild convulsions are really the signs of approaching dissolution.

CHAPTER X.

In order to judge of the present status of Darwinism it is of primary importance to note the position assumed by the few really eminent investigators, who as pupils of Haeckel still seem to have remained true to him. Among these I reckon Oskar Hertwig, the well known Berlin anatomist.

As early as 1899 in an address at the University on, Die Lehre vom Organismus und ihre Beziehung zur Sozialwissenschaft, Hertwig gave expression to views which are very little in harmony with the doctrines proceeding from Jena, and which are also put forth in his manual, The Cell and the Tissue. In that address we read (p. 8): "With the same right, with which, for the good of scientific progress, an energetic protest has been raised against a certain mysticism which attaches to the word Vitality, I beg to give warning against an opposite extreme which is but too apt to lead to onesided and unreal, and hence also, ultimately to false notions of the vital process, against an extreme which would see in the vital process nothing but a chemico-physical and mechanical problem and thinks to arrive at true scientific knowledge only in so far as it succeeds in tracing back phenomena to the movements of repelling and attracting atoms and in subjecting them to mathematical calculation."

With right does the physicist Mach, with reference to such views and tendencies, speak of a 'mechanical mythology in opposition to the animistic mythology of the old religions' and considers both as "improper and fantastic exaggerations based on a one-sided judgment." "My position on the question just stated becomes apparent from the consideration that the living organism is not only a complex of chemical materials and a bearer of physical forces, but also possesses a special organization, a structure, by means of which it is very essentially differentiated from the inorganic world, and in virtue of which it alone is designated as living."

Here, then, the distinction between living and non-living nature is clearly and definitely expressed, and Hertwig expresses himself just as definitely when he says (p. 21): "Whereas, but a few decades ago a scientific materialistic conception of the world issuing from a onesided, unhistorical point of view, misjudged the significance of the historic religious and ethical forces in the development of mankind, a change has become apparent in this regard."

To this gratifying testimony against materialism the distinguished naturalist added an equally valuable testimony regarding Darwinism on the occasion of the naturalists' convention in 1900. He there sketched an excellent summary of the "Development of Biology in the Nineteenth Century," in which he decidedly opposes the materialistic-mechanical conception of life. In so doing he also touches upon Haeckel's carbon-hypothesis, to which the latter still clings, and says: "That from the properties of carbon, combined with the properties of oxygen, hydrogen, nitrogen, etc., in certain proportions albumen should result, is a process which in its essence is as incomprehensible as that a living cell should arise from a certain organization of different albumina." Then the speaker is inevitably led to speak of the doctrine of Descent and Darwinism.

In the first place he declares definitely that ontogeny alone, i.e., the development of the individual being, is "capable of a direct scientific investigation." On the other hand we move in the domain of hypotheses in dealing with the further question: "How have the species of organisms living to-day originated in the course of the world's history?" This is a very valuable admission in view of Haeckel's dogmatic assertion that the descent of man from the ape is a "certain historical fact." Very moderate and pertinent are also the further words of the speaker: "Of course, a philosophically trained investigator will regard it as axiomatic that the organisms which inhabit our earth to-day did not exist in their present form in earlier periods of the earth and that they had to pass through a process of development, beginning with the simplest forms."

"But in the attempt to outline in detail the particular form in which a species of animals of our day existed in remote antiquity, we lose the safe ground of experience. For out of the countless millions of organisms, that lived in earlier periods of the earth, the duration of which is measured by millions of years, only scanty skeleton remains have by way of exception been preserved in a fossil state. From these naturally but a very imperfect and hypothetical representation can be formed of the soft bodies with which they were once clothed. And even then it remains forever doubtful whether the progeny of the prehistoric creature, the scant remains of which we study, has not become entirely extinct, so that it can in no way be regarded as the progenitor of any creature living at present." I should like to know wherein this differs radically from Fleischmann's contention in his "Descendenztheorie" (p. 10.) For we find stated here what Fleischmann emphasizes so much, viz., that with the problem of Descent we leave the domain of experience. It is worthy of special note in this connection that Hertwig likewise evidently regards as the sole really empirically and inductively serviceable proof of Descent, that which is drawn from palaeontology, from prehistoric animal and plant remains. He makes not the least mention of the indirect proofs taken from ontogenetic development or comparative anatomy, to which the Darwinians and advocates of Descent love so much to appeal, because they feel that the real inductive proof is lacking and totally fails to sustain their position. Hertwig next points out that the problem of Descent stirred scientific as well as lay circles twice during the past century. He then pays Lamarck and Darwin the necessary tribute, at which we cannot take offense since he was reared in the Darwinian atmosphere of Jena. I also willingly admit that Darwinism served science as a "powerful ferment," even if I must emphasize just as decidedly how harmful it was that this "ferment" was introduced into lay circles at an unseasonable time by the apostles of materialism. For while it was very well adapted to bring about in educated circles a fermentation which produced beneficial results, in uncritical lay-circles this ferment produced nothing but a corruption of world-views.

Hertwig then designates "Struggle for Existence," Survival of the Fittest, and Selection, as "very indefinite expressions." "With too general terms, one does not explain the individual case or produces only the appearance of an explanation whereas in every case the true causative relations remain in the dark. But it is the duty of scientific investigation to establish for each observed effect the prevenient cause, or more correctly, since nothing results from a single cause, to discover the various causes."

"The origin of the world of organisms from natural causes, however, is certainly an unusually complicated and difficult problem. It is just as little capable of being solved by a single magic formula as every disease is of yielding to a panacea. By the very act of proclaiming the omnipotence of natural selection, Weismann found he was forced to the admission that: "as a rule we cannot furnish the proof that a definite adaptation has originated through natural selection," in other words: We know nothing in reality of the complexity of causes which has produced the given phenomenon. So we may on the contrary, with Spencer, speak of the "Impotence of Natural Selection.""

"In this scientific struggle with which the past century closed, it seems necessary to distinguish between the doctrine of evolution and the theory of selection. They are based on entirely different principles. For with Huxley we can say: "Even if the Darwinian hypothesis were blown away, the doctrine of Evolution would remain standing where it stood." In it we possess an acquisition of our century which rests on facts, and which undoubtedly ranks amongst its greatest."

This last sentence affirms exactly what I have repeatedly asserted: the doctrine of Descent remains, Darwinism passes away. Hertwig then is decidedly of opinion that Darwinism entirely fails in the individual case because in its application the basis of experience vanishes. Indeed, according to him, phylogeny is not at all capable of direct scientific investigation. These are all important admissions which one would certainly have considered impossible twenty years ago; they unequivocally indicate the decline of Darwinian views, and in a certain way also harmonize with Fleischmann's work.

True, Hertwig still clings to the thought of Descent, but apparently no longer as to a conclusion of natural science. This appears from the assertion: "Ontogeny alone is capable of a direct scientific (he evidently speaks of natural science) investigation," and from the other statement that a philosophically trained investigator will accept it (Descent) as axiomatic although it belongs to the domain of hypothesis. What else does this mean but that: We have no specific knowledge of Descent but we believe in it. In short, this is not natural science but natural philosophy; it forms no constituent part of our certain knowledge of nature but it is one aspect of our world-view.

All the above-quoted assertions of Hertwig are calm and well-considered and show a decided deviation from the Darwinian position. Above all we are pleased to note that he appropriates Spencer's phrase regarding the "Impotence of Natural Selection" and that in the citation from Huxley he at least admits the possibility that the Darwinian doctrine will be "wafted away."

It is also proper to mention here the fact that in another place Hertwig no longer recognizes so fully the dogma set up by Fritz Mueller and Haeckel which is so closely bound up with Darwinism. I mean the so-called "biogenetic principle" according to which the individual organism is supposed to repeat in its development the development of the race during the course of ages.

In his book: "The Cell and the Tissue" (Die Zelle und die Gewebe, II. Jena 1898, p. 273) Hertwig says: "We must drop the expression: 'repetition of forms of extinct ancestors' and employ instead: repetition of forms which accord with the laws of organic development and lead from the simple to the complex. We must lay special emphasis on the point that in the embryonic forms even as in the developed animal forms general laws of the development of the organized body-substance find expression."

Any one can subscribe to these statements; in truth they contain something totally different from the "biogenetic principle"; for Haeckel has really no interest in so general a truth, but is intent only upon a proof of Descent.

Hertwig continues: "In order to make our train of thought clear, let us take the egg-cell. Since the development of every organism begins with it, the primitive condition is in no way recapitulated from the time when perhaps only single-celled amoebas existed on our planet. For according to our theory the egg-cell, for instance, of a now extant mammal is no simple and indifferent, purposeless structure, as it is often represented, (as according to Haeckel's "biogenetic principle" it would necessarily be); we see in it, in fact, the extraordinarily complex end-product of a very long historic process of development, through which the organic substance has passed since that hypothetical epoch of single-celled organisms."

"If the eggs of a mammal now differ very essentially from those of a reptile and of an amphibian because in their organization they represent the beginnings only of mammals, even as these represent only the beginnings of reptiles and amphibians, by how much more must they differ from those hypothetical single-celled amoebas which could as yet show no other characteristics than to reproduce amoebas of their own kind."

This is a view which has frequently been clearly expressed by anti-Darwinians: The egg-cells of the various animals are in themselves fundamentally different and can therefore have nothing in common but similarity of structure. In opposition to Hertwig, Haeckel in his superficial way deduces from it an internal similarity as well. After a few polite bows before his old teacher, Haeckel, Hertwig thus summarizes his view: "Ontogenetic (that is, those stages in the individual development) stages therefore give us only a greatly changed picture of the phylogenetic (i.e., genealogical) stages as they may once have existed in primitive ages, but do not correspond to them in their actual content." This is a very resigned position, very far removed from Haeckel's certainty and orthodoxy.

To sum up: O. Hertwig has become a serious heretic in matters Darwinian. Will Haeckel, in his usual manner try to cast suspicion on Hertwig also? For Haeckel himself says (Free Science and Free Doctrine, Stuttgart, 1878, p. 85): "Since I am not bound by fear to the Berlin Tribunal of Science or by anxieties regarding the loss of influential Berlin connections, as are most of my like-minded colleagues, I do not hesitate here as elsewhere to express my honest conviction, frankly and freely, regardless of the anger which perhaps real or pretended privy councillors in Berlin may feel upon hearing the unadorned truth."

Verily, it is a matter of suspense to know whether his school will now pour forth their wrath upon O. Hertwig, or whether finally the discovery will not be made in Jena that Hertwig secretly possessed himself of his position in Berlin, in the same manner as Fleischmann obtained his at Erlangen, viz., by a promise of desertion from Darwinism.

CONCLUSION.

We may conveniently summarize what we have said in the foregoing chapters in the following statement: The theory of Descent is almost universally recognized to-day by naturalists as a working hypothesis. Still, in spite of assertions to the contrary, no conclusive proof of it has as yet been forthcoming. Nevertheless it cannot be denied that the theory provides us with an intelligible explanation of a series of problems and facts which cannot be so well explained on other grounds.

On the other hand, Darwinism, i.e., the theory of Natural Selection by means of the Struggle for Existence, is being pushed to the wall all along the line. The bulk of naturalists no longer recognizes its validity, and even those who have not yet entirely discarded it, are at least forced to admit that the Darwinian explanation now possesses a very subordinate significance.

In the place of Darwinian principles, new ideas are gradually winning general acceptance, which, while they are in harmony with the principles of adaptation and use, (Lamarck) enunciated before the time of Darwin, nevertheless attribute a far-reaching importance to internal forces of development. These new conceptions necessarily involve the admission that Evolution has not been a purely mechanical process.

THE BOOK OF THE DAY

Science and Christianity

By F. BETTEX

Translated from the German

The author among other things says in the preface: I wish to make clear to my readers how little real science is hidden behind the fine phrases and sounding words or the infidel, and how little he himself understands of the material creation which he affirms to be the only one.... The Christian and Biblical conception of the universe is more logical, more harmonious, more in accordance with facts, therefore, more scientific than all philosophies, all systems, materialistic and atheistic. Contents of the book:

Chapter I.

Progress

Chapter II.

Evolution and Modern Science

Chapter III.

Christians and Science

Chapter IV.

Science

Chapter V.

Materialism

One of the many favorable reviews: It is a view of much scope, and so far as it attempts reconciliation between science and christianity, is eminently successful. There can be no doubt that at present, when there is so pronounced a disposition to follow every fad in science, especially if it opposes the Bible, such a book should have a wide reading and is adapted to accomplish much good.

Price $1.50

GERMAN LITERARY BOARD, Burlington, Iowa

End of Project Gutenberg's At the Deathbed of Darwinism, by Eberhard Dennert

*** END OF THIS PROJECT GUTENBERG EBOOK AT THE DEATHBED OF DARWINISM ***

***** This file should be named 21019.txt or 21019.zip ***** This and all associated files of various formats will be found in: http://www.gutenberg.org/2/1/0/1/21019/

Produced by Bryan Ness, Jamie Atiga and the Online Distributed Proofreading Team at http://www.pgdp.ne

Updated editions will replace the previous one--the old editions will be renamed.

Creating the works from public domain print editions means that no one owns a United States copyright in these works, so the Foundation (and you!)

can copy and distribute it in the United States without permission and without paying copyright royalties. Special rules, set forth in the General Terms of Use part of this license, apply to copying and distributing Project Gutenberg-tm electronic works to protect the PROJECT GUTENBERG-tm concept and trademark. Project Gutenberg is a registered trademark, and may not be used if you charge for the eBooks, unless you receive specific permission. If you do not charge anything for copies of this eBook, complying with the rules is very easy. You may use this eBook for nearly any purpose such as creation of derivative works, reports, performances and research. They may be modified and printed and given away--you may do practically ANYTHING with public domain eBooks. Redistribution is subject to the trademark license, especially commercial redistribution.

###

www.ingramcontent.com/pod-product-compliance
Lightning Source LLC
Chambersburg PA
CBHW051814170526
45167CB00005B/2014